Proceedings in Engineering Mechanics

Research, Technology and Education

Series Editors

This book series publishes the results of meetings dealing with material properties in engineering and science. It covers a wide range of topics, from the fundamentals of materials mechanics and applications for various industries to aspects of scientific training and career development. The volumes in the series are based typically on primary research materials presented at conferences, workshops, and similar scientific meetings, and represent comprehensive scientific and technical studies.

Lucas F. M. da Silva · Robert D. Adams
Editors

8th International Conference on Structural Adhesive Bonding 2025

Selected Contributions of AB 2025

Springer

Editors
Lucas F. M. da Silva
Faculty of Engineering
University of Porto
Porto, Portugal

Robert D. Adams
Oxford, UK

ISSN 2731-0221 ISSN 2731-023X (electronic)
Proceedings in Engineering Mechanics
ISBN 978-3-032-19021-5 ISBN 978-3-032-19022-2 (eBook)
https://doi.org/10.1007/978-3-032-19022-2

This Springer imprint is published by the registered company Springer Nature Switzerland AG
The registered company address is: Gewerbestrasse 11, 6330 Cham, Switzerland

Preface

This volume of the series *Proceedings in Engineering Mechanics-Research, Technology and Education* contains selected papers presented at the 8th International Conference on Structural Adhesive Bonding (AB 2025), held in Porto (Portugal) during 10–11 July 2025 (https://www.engeduconferences.com/AB2025/). The goal of the conference was to provide a unique opportunity to exchange information, present the latest results as well as to discuss issues relevant to structural adhesive bonding research today. Approximately 160 papers were presented by researchers from more than 20 countries.

In order to disseminate the work presented in AB 2025, selected papers were prepared which resulted in the present volume dedicated to adhesive bonding. A wide range of topics are covered resulting in 12 papers dealing with the most recent research topics concerning adhesive bonding: adhesive formulation and properties (first four papers), adhesion and surface treatments (two papers), manufacturing (two papers), joint design (one paper) and applications (last three papers). The book represents the latest trends in adhesive joining and serves as a reference volume for researchers and graduate students working with adhesive bonding.

The organizer and editor wish to thank all the authors for their participation and cooperation, which made this volume possible. Finally, I would like to thank the team of Springer-Verlag, especially Dr. Christoph Baumann and Ute Heuser, for the excellent cooperation during the preparation of this volume.

September 2025

Lucas F. M. da Silva
Robert D Adams

Contents

Joint design

Applications

Adhesive Formulation and Properties

Metal Oxide Fillers in UV-Resistant Adhesives: Analysis of Selected Compositions

Izabela Miturska-Barańska[1](✉), Anna Rudawska[1], Miroslav Müller[2], and Dana Stančeková[3]

[1] Faculty of Mechanical Engineering, Lublin University of Technology, Nadbystrzycka 36, 20-618 Lublin, Poland
{i.miturska,a.rudawska}@pollub.pl

[2] Faculty of Engineering, Czech University of Life Sciences Prague, Kamycka 129, 165 00 Prague, Czech Republic
muller@tf.czu.cz

[3] Faculty of Mechanical Engineering, University of Žilina, Univerzitná 1, 010 26 Žilina, Slovakia
Dana.Stancekova@fstroj.uniza.sk

Abstract. The stability of adhesive joints under ultraviolet (UV) radiation is a critical issue in automotive, aerospace, electronics, and construction applications. UV-induced degradation leads to bond scission, oxidation, embrittlement, and yellowing of polymer matrices, which significantly reduces adhesive performance. One of the most effective approaches to improve UV resistance is the incorporation of inorganic fillers, particularly metal oxides. This chapter provides a review of the role of selected metal oxides (TiO_2, Al_2O_3, ZnO, SiO_2, Fe_2O_3) in adhesives and polymer systems, focusing on their mechanisms of action, advantages, and limitations. A case study is also presented, in which an epoxy adhesive system (L + CL) was modified with TiO_2 and Al_2O_3 at 1 and 3 wt.% loadings. Samples were exposed to accelerated UV aging (24 h, 72 h, 240 h) and evaluated in terms of colour change and tensile strength retention. The results confirmed that TiO_2, especially at 3 wt.%, effectively reduced yellowing and improved mechanical stability, while Al_2O_3 showed limited protective action. These findings highlight the potential of metal oxide fillers for designing UV-resistant adhesive systems and point to future directions for long-term durability studies.

Keywords: metal oxide fillers · UV-resistant adhesives · titanium dioxide (TiO_2) · aluminium oxide (Al_2O_3) · epoxy adhesives · accelerated UV aging · colour stability · tensile strength

1 Introduction

Structural and specialty adhesives have become key components in modern engineering applications, including the automotive, aerospace, electronics, machinery, and packaging industries. In many of these areas, adhesive joints are exposed to ultraviolet (UV) radiation, often in combination with fluctuating temperature and humidity. It is well

L. F. M. da Silva and R. D. Adams (Eds.): AB 2025, PEM, pp. 3–18, 2026.
https://doi.org/10.1007/978-3-032-19022-2_1

established that UV exposure initiates bond scission, free radical generation, and photo-oxidation in polymers, ultimately leading to reduced mechanical strength, embrittlement, and yellowing of the adhesive matrix, and in the long term, to joint degradation (Yousif and Haddad, 2013; Miturska et al., 2020; Chamas et al., 2020; Andrady et al., 2022).

One of the most effective approaches to improving UV stability is the incorporation of inorganic fillers—particularly metal oxides (e.g., TiO_2, Al_2O_3, ZnO, SiO_2). Titanium dioxide is both the most widely used white pigment and an optically active material, whose interaction with light (UV absorption and scattering) can significantly reduce photon penetration into the polymer matrix. In polymer systems, TiO_2 therefore acts as a UV "screen" although its effectiveness depends on crystal structure, particle size, degree of agglomeration, and surface modification (Chen and Mao, 2007; Nakata and Fujishima, 2012).

It is important to emphasize the ambivalent character of TiO_2: at the nanoscale and with an "active" surface it may accelerate photodegradation through radical catalysis, while in passivated forms (e.g., rutile coated with SiO_2/Al_2O_3) it acts primarily as an optical barrier and colour stabilizer (Egerton, 2013; Affrossman et al., 2015). Critical parameters include particle size, surface treatment, and volume fraction within the matrix. Numerous studies confirm that smaller particles increase the effective surface area for scattering and absorption, but simultaneously raise the risk of photocatalytic activity if the surface is not properly modified (Wojciechowski et al., 2015; Chen et al., 2023).

Aluminum oxide (Al_2O_3) represents another widely used filler. Unlike TiO_2, Al_2O_3 exhibits no significant photocatalytic activity; its primary role is to enhance the polymer matrix through barrier and reinforcing effects (e.g., increased stiffness, reduced oxygen/water diffusion, modified interfacial contact), which translate into improved aging stability (Petković et al., 2019; Petković et al., 2024). In numerous polymer composites, including epoxy and silicone systems, the addition of Al_2O_3 increases modulus and hardness and supports the retention of properties after environmental exposure (Korkmaz and Gültekin, 2022).

Within the field of adhesives, two groups of observations are of particular importance: the effect of filler type on the strength and durability of joints after aging cycles (UV and hygrothermal), and the influence of modification on optical properties and aesthetics (yellowing, gloss loss). For PVAc adhesives, it has been shown that SiO_2/TiO_2 nanoparticles improve elasticity and aging resistance (Petković et al., 2019; Petković et al., 2024), while in epoxy adhesives, nanofillers such as TiO_2, h-BN, and B_4C enhance performance after UV conditioning (Korkmaz and Gültekin, 2022). Furthermore, studies on adhesive joints have demonstrated that UV exposure significantly reduces mechanical–adhesive strength, whereas the use of inorganic stabilizers can mitigate this effect (Amorim et al., 2018).

Research on coatings and polymer composites—materials chemically and functionally related to adhesives—confirms that properly engineered TiO_2 particles reduce colour change, gloss loss, and mechanical deterioration during accelerated weathering. However, excessive loading or inadequate surface modification of the pigment may hinder crosslinking or induce coating cracking (Wojciechowski et al., 2015; Pan et al., 2021; Chen et al., 2023).

Despite extensive literature on UV stabilization of polymer coatings and composites, comparative analyses of adhesives modified with different metal oxides under identical aging conditions and comparable filler loadings remain limited. Therefore, the present chapter provides an overview of the mechanisms of interaction between metal oxides and polymer matrices under UV exposure, with particular emphasis on TiO_2 and Al_2O_3, as well as a case study based on experimental research. In this study, an epoxy adhesive system (L + CL) was modified with TiO_2 and Al_2O_3 at loadings of 1 and 3 wt.%. Samples were UV-aged for 24 h, 72 h, and 240 h, followed by evaluation of tensile strength and visual assessment of colour changes (Wei et al., 2024).

2 Metal Oxides as Fillers in Adhesive Compositions – Literature Review

The incorporation of metal oxides into polymeric and adhesive materials has been shown to significantly modify both their mechanical performance and their resistance to environmental factors. The most widely studied fillers include titanium dioxide (TiO_2), aluminium oxide (Al_2O_3), zinc oxide (ZnO), silicon dioxide (SiO_2), and iron oxides (Fe_2O_3). Each of these materials interacts differently with the polymer matrix and provides a distinct potential for protection against ultraviolet (UV) radiation.

TiO_2 is the best-known and most commonly applied mineral filler, valued for its ability to absorb and scatter UV radiation. Its effectiveness as a stabilizer depends strongly on particle size, crystal phase, and the type of surface passivation applied. Rutile forms coated with SiO_2 or Al_2O_3 exhibit reduced photocatalytic activity, thereby protecting polymers against degradation while minimizing the formation of reactive oxygen species (Chen and Mao, 2007; Nakata and Fujishima, 2012; Egerton, 2013; Zeljko et al., 2021; Kim et al., 2023). By contrast, Al_2O_3 does not exhibit photocatalytic reactivity and mainly acts as a reinforcing and barrier filler. The incorporation of Al_2O_3 increases stiffness, hardness, and mechanical durability of adhesive joints, while also limiting oxygen and moisture diffusion into the bulk polymer, which slows down the rate of degradation (Abdeen et al., 2019; Zhang et al., 2025).

ZnO also plays an important role, as it efficiently absorbs UV radiation in a manner similar to TiO_2. However, ZnO requires suitable surface modification to avoid photocatalytic side reactions and agglomeration tendencies (Şomoghi et al., 2024). SiO_2 enhances hardness, stiffness, and abrasion resistance and, when combined with organic stabilizers, exhibits synergistic effects in protecting against aging (Nguyen et al., 2020; Martins et al., 2021). Iron oxides, such as Fe_2O_3, act primarily as UV absorbers and simultaneously improve corrosion resistance in coatings, though their use entails changes in colour and requires careful control of filler content (Dhoke and Khanna, 2009). Figure 1 illustrates the principal protective mechanisms of metal oxides in adhesive compositions.

These mechanisms include optical effects (UV absorption and scattering, reducing photon penetration into polymer bonds), barrier effects (limiting gas and moisture permeability), and mechanical effects (improving modulus and strength through well-dispersed particles). Without proper modification, however, some oxides (particularly TiO_2 and ZnO) may act photocatalytically, accelerating polymer degradation (Egerton,

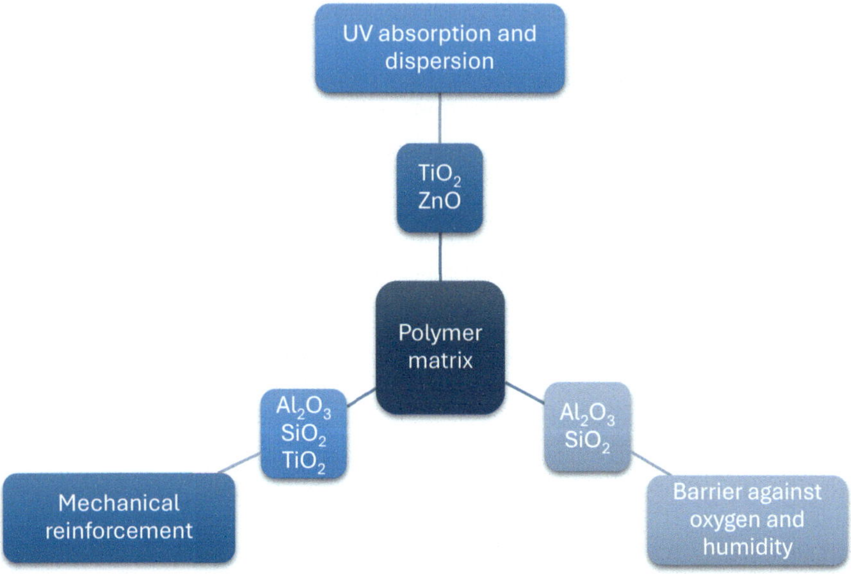

Fig. 1. Protective mechanisms of metal oxides in adhesive compositions

Table 1. Examples of metal oxide fillers in polymer systems [compiled from cited sources]

Filler	Matrix	Effect	Reference
TiO_2	PVAc	improved adhesion, reduced degradation	(Petković et al., 2019; Petković et al., 2024)
TiO_2 + h-BN, B_4C	epoxy	higher strength retention after UV exposure	(Korkmaz and Gültekin, 2022)
ZnO	epoxy	enhanced mechanical and UV resistance	(Şomoghi et al., 2024)
SiO_2	PU-acrylic	synergistic effect with organic stabilizers	(Nguyen et al., 2020; Martins et al., 2021)
Fe_2O_3	alkyd	UV and corrosion resistance, colour change	(Dhoke and Khanna, 2009)

2013; Kim et al., 2023). Examples of filler applications in different polymer systems are summarized in Table 1.

The impact of UV radiation on adhesives has been extensively described in the literature. The aging process involves bond scission, reduction in molecular weight, microcrack formation, and loss of cohesion, which lead to decreased mechanical strength and discoloration (Yousif and Haddad, 2013; Andrady et al., 2022). For adhesive joints, UV exposure significantly reduces shear and tensile strength and alters modulus (Amorim et al., 2018). The addition of inorganic fillers can substantially mitigate these adverse effects.

Numerous studies on adhesive and coating systems confirm this tendency. For PVAc adhesives, the addition of SiO_2 and TiO_2 nanoparticles improves adhesion and limits degradation during accelerated aging (Petković et al., 2019; Petković et al., 2024). In epoxy adhesives, nanofillers such as TiO_2, h-BN, or B_4C increase strength retention after UV exposure, stabilizing adhesive joints. For instance, Korkmaz and Gültekin (Korkmaz and Gültekin, 2022) demonstrated that epoxy adhesives reinforced with 2 wt.% h-BN or B_4C exhibited higher mechanical retention after 480–1440 h of continuous UV exposure compared to unmodified references. Both bulk adhesives and aluminium–epoxy adhesive joints were tested, and tensile performance consistently decreased with exposure time, though the reduction was less severe in modified systems. Dynamic mechanical analysis confirmed that nanofillers limited chain mobility and stabilized the adhesive structure under UV conditions.

Similarly, epoxy composites with h-BN and B_4C retained higher mechanical properties after UV aging compared to unmodified systems (Gültekin et al., 2021). Surface-modified nano-TiO_2 also showed beneficial effects on mechanical stability and morphology of epoxy composites (Goyat et al., 2018; Xia et al., 2023). Fernández-Álvarez et al. (Fernández-Álvarez et al., 2021) reported improved hardness (HU) and plastic work (W_plast) of epoxy coatings after 500 h of UV exposure when nano-SiO_2 was incorporated, with 3 wt.% SiO_2 yielding even higher values than the non-exposed reference (Fig. 2). These findings indicate that SiO_2 nanoparticles help retain mechanical properties of epoxy matrices under UV aging.

Based on the presented results, it was observed that for unfilled epoxy coatings hardness (HU) decreased compared to non-exposed samples. The addition of nano-SiO_2 partially compensated for this effect, and at a loading of 3 wt.% the hardness after 500 h of UV exposure was even higher than that of the non-exposed reference (AR). Similar trends were recorded for plastic work (W_plast), which stabilized or slightly increased upon incorporation of the filler. These findings indicate that SiO_2 nanoparticles enhance the retention of mechanical properties of epoxy coatings under UV radiation. In particular, the addition of 3 wt.% SiO_2 effectively limited the degradation of mechanical performance compared to the reference samples.

In UV-cured coatings, silica-coated TiO_2 was shown to effectively reduce yellowing and gloss loss while minimizing photocatalytic degradation risks (Zeljko et al., 2021; Chen et al., 2023). In epoxy composites, ZnO improved both UV and mechanical resistance, provided good dispersion and surface modification were ensured (Şomoghi et al., 2024). SiO_2-based systems exhibited synergistic stabilization when combined with organic stabilizers, maintaining high transparency along with enhanced UV durability (Nguyen et al., 2020; Martins et al., 2021). Finally, Fe_2O_3 additions in alkyd coatings improved UV and corrosion resistance, though adjustments in dosage were necessary to manage colour change (Dhoke and Khanna, 2009).

In summary, the literature clearly demonstrates that metal oxides, depending on type, surface treatment, and content, can significantly improve UV stability of adhesives. The greatest benefits are observed at low to moderate loadings (0.5–3 wt.%), provided good dispersion and compatibility with the polymer matrix are achieved.

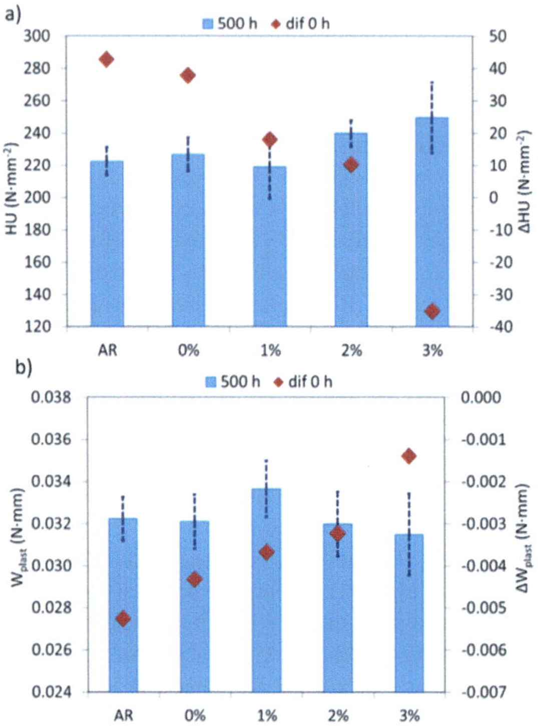

Fig. 2. Mechanical properties (HU and Wplast) of the epoxy coatings and the difference with respect to non-exposed coatings (Fernández-Álvarez et al., 2021)

3 Materials and Methods

A two-component epoxy system consisting of resin L and curing agent CL (R&G Faserverbundwerkstoffe GmbH, Germany) was used in this study. Resin L is an unpigmented, medium-viscosity bisphenol A/F-based epoxy resin with an epoxy group content of 5.0–5.3 eq/kg and a viscosity of 1100–1400 mPa·s (25 °C). Curing agent CL is a mixture of aliphatic amines with medium reactivity, solvent-free, applied at a weight ratio of 100:30 (resin: curing agent). The pot life of the mixture was approximately 60 min, and full curing occurred after 24 h at room temperature.

To modify the formulations, two types of inorganic fillers were employed: aluminium oxide (Al_2O_3) and titanium dioxide (TiO_2), supplied by WARCHEM Sp. z o.o. (Warsaw, Poland). Aluminum oxide (Al_2O_3, molar mass 101.94 g/mol) had a purity of $\geq$ 99.5% and a particle size of 1–10 μm, while titanium dioxide (TiO_2, molar mass 79.9 g/mol) exhibited a purity of $\geq$ 99.9% and a particle size in the micrometre range. Both fillers were incorporated into the resin at 1 wt.% and 3 wt.% relative to the resin mass. The composition of the tested systems is summarized in Table 2.

Sample preparation included: weighing of components, dispersion of fillers in the resin using a mechanical stirrer (460 rpm, 2 min), addition of the curing agent followed by further mixing, degassing under reduced pressure (approx. 500 mbar, 2 min), casting

Table 2. Composition of adhesive formulations

Composition designation	Filler type	Content [wt.%]	Notes
L/CL/100:30	–	0	Reference sample
L/CL/TiO_2/100:30:1	TiO_2	1	–
L/CL/ TiO_2/100:30:3	TiO_2	3	–
L/CL/Al_2O_3/100:30:1	Al_2O_3	1	–
L/CL/Al_2O_3/100:30:3	Al_2O_3	3	–

into silicone moulds, and curing at room temperature for 5 days. The entire procedure is schematically illustrated in Fig. 3.

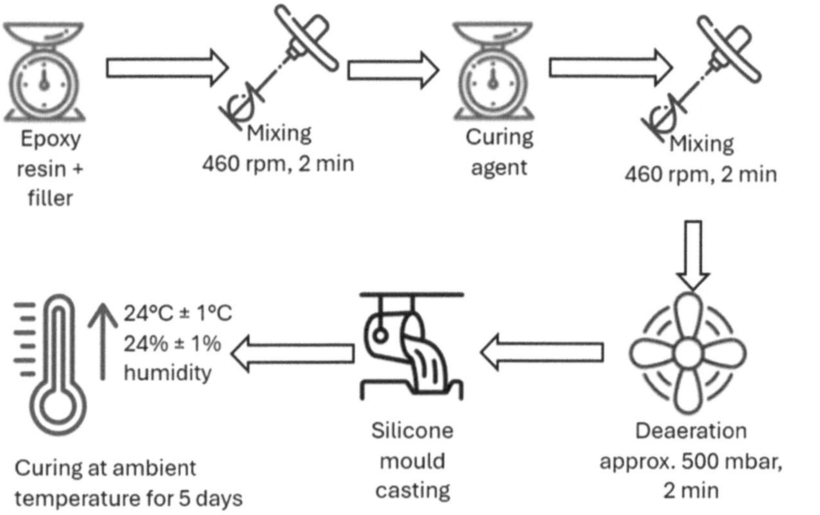

Fig. 3. Schematic representation of the sample preparation process

The geometry of the cured specimens used for testing is shown in Fig. 4.

The cured samples were subjected to UV aging in a Q-SUN Xe-3-HS chamber (Q-Lab, Germany) equipped with XeLamps X1800 xenon lamps. The exposure was performed at a wavelength of 320 nm, temperature of 40 °C, and relative humidity of 40%. Exposure times were 24, 72, and 240 h. The aging parameters are summarized in Table 3.

Mechanical tensile strength and visual colour changes of the samples were evaluated. Tensile tests were carried out in accordance with PN-EN ISO 527–1 using a Shimadzu EZ-Test-LX testing machine (Shimadzu Corp., Japan) with a 5 kN load cell. Each material variant was tested on five specimens. Visual colour assessment was performed under standardized lighting conditions (daylight D65) against a white background, with progressive changes documented photographically.

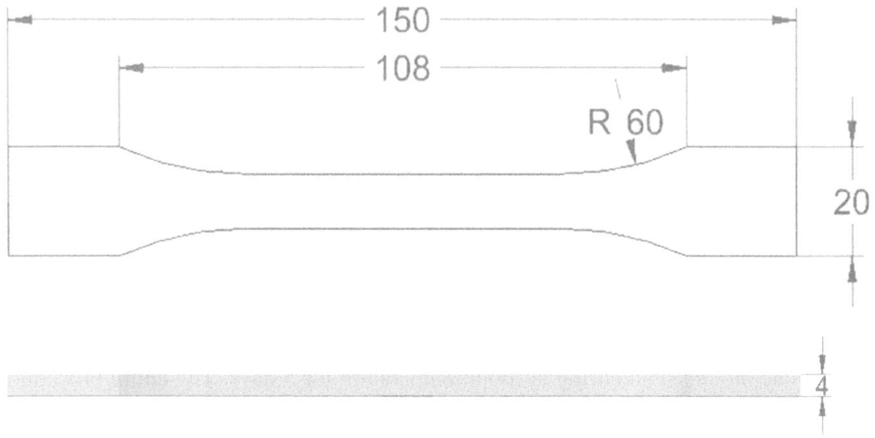

Fig. 4. Adhesive specimens prepared for testing

Table 3. Parameters of UV aging conditions for tested samples

Sample code	Curing time [h]	UV exposure time [h]	UV exposure temperature [°C]	RH [%]
Ref	120	0	-	-
24h	120	24	40	40
72h	120	72	40	40
240h	120	240	40	40

4 Results

The following section presents the results of tests performed on epoxy compositions modified with metal oxides (Al_2O_3, TiO_2) at different loadings (1 and 3 wt.%). Two aspects were analysed: visual changes in specimen colour during UV aging and changes in mechanical performance, expressed as tensile strength in accordance with PN-EN ISO 527–1.

The selection of these parameters was motivated by the fact that the primary effects of epoxy photodegradation are:

- optical changes (yellowing, haze), associated with oxidation processes and the formation of chromophores (Nikafshar et al., 2017; Ci et al., 2025),
- loss of mechanical strength, caused by bond scission, molecular weight reduction, and microcrack formation (Andrady et al., 2023).

Literature reports indicate that inorganic nanofillers can mitigate such phenomena, either through their ability to absorb and scatter UV radiation (TiO_2, ZnO) or through barrier effects (Al_2O_3, SiO_2) (Chen et al., 2023). Therefore, it was expected that modification with TiO_2 and Al_2O_3 would improve property retention during prolonged UV exposure.

4.1 Colour Changes After UV Exposure

After each UV exposure interval, a visual assessment of the specimen surfaces was conducted. The results are shown in Fig. 5.

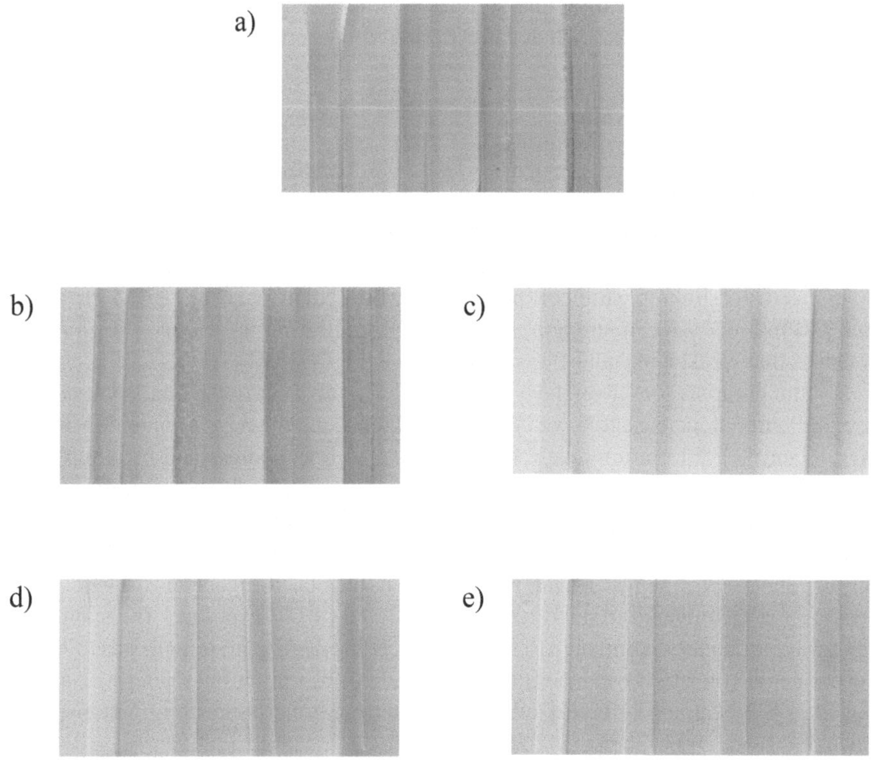

Fig. 5. Visual evaluation of epoxy adhesive specimens: a) L/CL/100:30, b) L/CL/Al_2O_3/100:30:1, c) L/CL/Al_2O_3/100:30:3, d) L/CL/TiO_2/100:30:1, e) L/CL/TiO_2/100:30:3. From left to right: reference, exposed for 24 h, 72 h, and 240 h.

For the reference composition (L/CL/100:30), containing no filler, initial yellowing was already visible after 24 h, which intensified with longer aging times. After 240 h, the specimens exhibited a pronounced amber coloration, consistent with the typical photodegradation mechanism of epoxy systems involving bond scission and oxidative product formation (Andrady et al., 2023; Ci et al., 2025).

The addition of Al_2O_3 at 1 wt.% (L/CL/Al_2O_3/100:30:1) only slightly delayed colour changes—after 72 h, samples already showed noticeable yellowing, and after 240 h their appearance was comparable to the reference. At 3 wt.% Al_2O_3, no stabilizing effect was observed; on the contrary, specimens darkened similarly to the reference, suggesting that the barrier effect of Al_2O_3 (Singla et al., 2014; Bazrgari et al., 2018) is insufficient to prevent visual photodegradation.

In contrast, samples containing TiO_2 behaved differently. At 1 wt.% TiO_2, colour changes were less pronounced compared to the reference, but the most beneficial results were observed at 3 wt.%. After 240 h of UV exposure, TiO_2-modified specimens remained noticeably lighter than both the reference and Al_2O_3-modified samples, confirming the ability of TiO_2 to absorb and scatter UV radiation (Chen and Mao, 2007; Nakata and Fujishima, 2012). However, literature emphasizes that stabilization efficiency depends on the crystalline form and surface treatment of the particles, as unmodified TiO_2 can in some cases exhibit photocatalytic activity (Egerton, 2013; Kim et al., 2023).

4.2 Tensile Strength Results

W Following visual inspection, tensile tests were carried out. Average results are presented in Figs. 6 and 7.

For the reference specimens (L/CL/100:30), tensile stress and maximum force initially increased after short-term UV exposure (24 h and 72 h), likely due to additional UV-induced crosslinking within the polymer network (Nikafshar et al., 2017). However, after 240 h, a decrease in strength was observed, indicating the onset of degradation processes that weaken the adhesive structure (Ci et al., 2025).

For the compositions with 1 wt.% Al_2O_3 (Figs. 6b and 7b), a similar trend was observed—initial increase followed by decline. Increasing the Al_2O_3 content to 3 wt.% (Figs. 6 and 7c) did not improve retention after 240 h UV; stress and breaking force values were comparable to the reference. These results suggest that the barrier effect of Al_2O_3 (Bazrgari et al., 2018) was insufficient to significantly limit degradation.

Conversely, TiO_2-modified samples showed better retention. Both at 1 wt.% (Figs. 6 and 7d) and 3 wt.% (Figs. 6 and 7e), specimens achieved higher tensile stress and breaking force compared to the reference after 240 h of UV exposure. The stabilizing effect was particularly pronounced at 3 wt.%, where the decline in strength was minimal. The protective role of TiO_2 can be attributed to its UV absorption and scattering capacity as well as a potential reinforcing effect associated with the presence of nanoparticles (Chen and Mao, 2007; Chen et al., 2023).

5 Discussion

The conducted study allows for a comprehensive evaluation of the effect of nanofillers in the form of Al_2O_3 and TiO_2 on the UV resistance of epoxy compositions. The results clearly demonstrate that both the type of filler and its loading determine the extent of visual changes (yellowing, darkening) as well as the retention of mechanical properties (tensile stress, breaking force).

In the reference samples, rapid yellowing was observed already after 24–72 h of UV exposure, and after 240 h their colour turned to an intense amber shade. The addition of Al_2O_3 did not result in significant improvement—at both 1% and 3% loadings, the rate of colour change was comparable to that of the unmodified material. The barrier

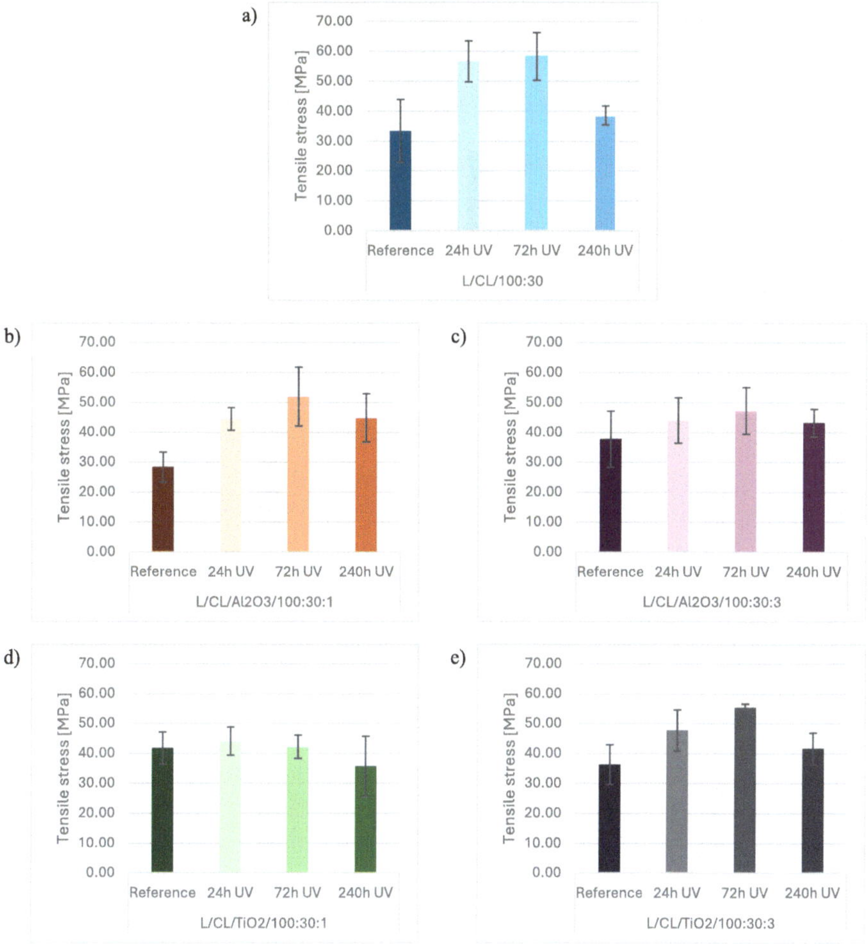

Fig. 6. Average tensile stresses of epoxy adhesive compositions before and after UV aging: a) L/CL/100:30, b) L/CL/Al_2O_3/100:30:1, c) L/CL/Al_2O_3/100:30:3, d) L/CL/TiO_2/100:30:1, e) L/CL/TiO_2/100:30:3

mechanism of Al_2O_3, associated with the reduction of oxygen and moisture diffusion (Bazrgari et al., 2018), proved insufficient to suppress the formation of oxidative products responsible for yellowing (Ci et al., 2025).

In contrast, specimens containing TiO_2, particularly at 3 wt.%, showed a clear reduction in discoloration. After 240 h, their colour was visibly lighter than that of the reference and Al_2O_3-modified samples. This effect can be attributed to the ability of TiO_2 to absorb and scatter UV radiation (Chen and Mao, 2007), thereby reducing the amount of energy reaching polymer bonds. Similar observations were made by Chen et al. (Chen et al., 2023) and Zeljko et al. (Zeljko et al., 2021), who confirmed that TiO_2 in polymer composites effectively limits yellowing and gloss loss during UV aging.

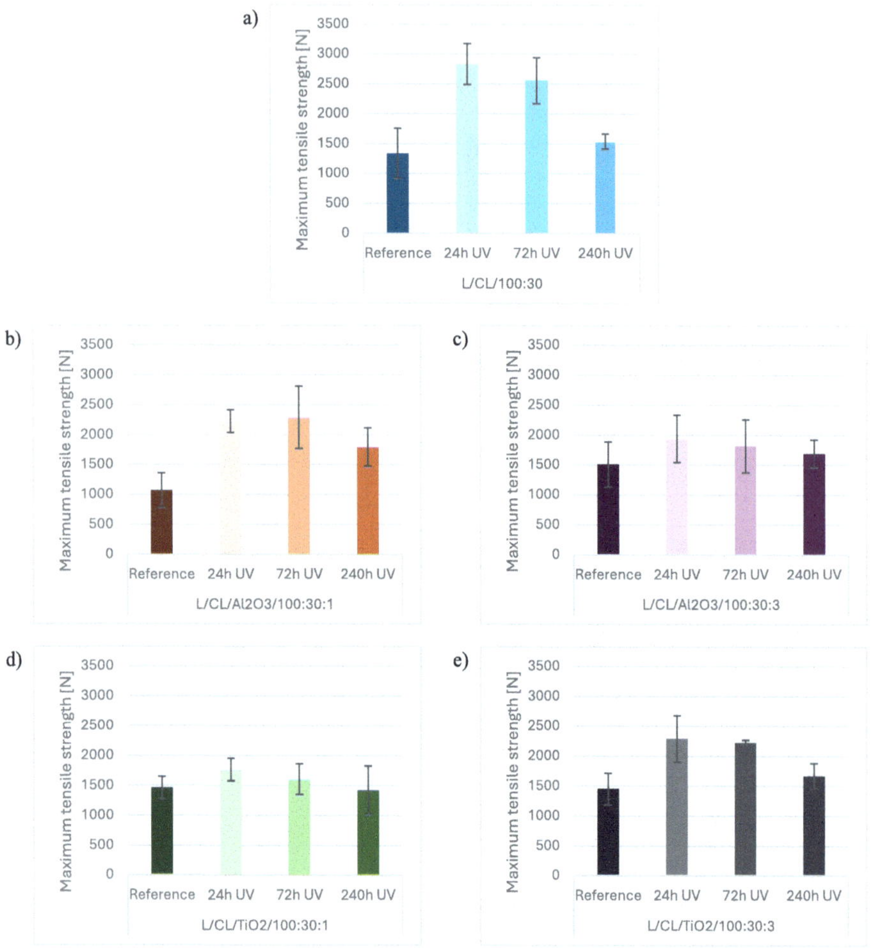

Fig. 7. Average maximum breaking force in tensile tests of epoxy adhesive compositions before and after UV aging: a) L/CL/100:30, b) L/CL/Al_2O_3/100:30:1, c) L/CL/Al_2O_3/100:30:3, d) L/CL/TiO_2/100:30:1, e) L/CL/TiO_2/100:30:3

Mechanical test results further corroborated these trends. Short-term UV exposure (24 h and 72 h) led to an initial increase in tensile stress and breaking force, which can be linked to additional crosslinking induced by radiation (Nikafshar et al., 2017). However, with prolonged exposure (240 h), all samples exhibited decreased strength, resulting from chain scission and microcrack formation (Ci et al., 2025).

For Al_2O_3-modified compositions, both at 1% and 3%, values were close to those of the reference specimens. This confirms that Al_2O_3 primarily acts as a mechanical reinforcing agent (Bazrgari et al., 2018), without providing effective protection against photodegradation.

Conversely, TiO_2-modified samples exhibited higher tensile stress and breaking force compared to the reference after 240 h of UV exposure. The best retention was achieved at 3 wt.% TiO_2, suggesting an optimal protective effect at this concentration. The stabilizing role of TiO_2 results not only from its optical shielding mechanisms but also from possible improvements in nanoscale adhesion through uniform particle dispersion (Chen and Mao, 2007; Nakata and Fujishima, 2012). Comparable findings were reported by Korkmaz and Gültekin (Korkmaz and Gültekin, 2022), who demonstrated improved strength retention of epoxy joints modified with nanofillers such as BN and B_4C under UV exposure.

The obtained results highlight significant differences in the mechanisms of the two tested oxides. TiO_2, owing to its optical properties, effectively suppresses both visual changes and mechanical degradation, whereas Al_2O_3 acts mainly through barrier and reinforcing effects, which do not translate into improved UV resistance under the studied conditions.

From a practical perspective, the most favourable performance was achieved at 3 wt.% TiO_2. This outcome is consistent with literature, where low to moderate nanofiller loadings (0.5–3 wt.%) are most often indicated as optimal for property retention (Chen and Mao, 2007; Chen et al., 2023).

It should be emphasized that the present study covered a relatively short UV exposure period (max. 240 h). Future work should include long-term testing under real environmental conditions (temperature fluctuations, humidity, day/night cycles). Complementary analytical techniques, such as FTIR, SEM, or DMA, would enable more detailed identification of degradation mechanisms and verification of filler roles in stabilizing epoxy systems.

6 Conclusion

The conducted study enabled the evaluation of the influence of selected metal oxides (Al_2O_3, TiO_2) on the UV resistance of L/CL epoxy compositions. Two key aspects were analysed: colour changes of the samples during UV aging and the retention of mechanical properties assessed through tensile strength measurements.

Based on the obtained results, the following conclusions can be drawn:

- Effect of UV on reference samples – compositions without fillers exhibited noticeable yellowing already after 24 h of exposure, which intensified over time. At the same time, the initial increase in mechanical strength (24–72 h) suggests additional radiation-induced crosslinking of the polymer network, whereas after 240 h a significant decrease in strength occurred due to degradation processes.
- Addition of Al_2O_3 – at both 1% and 3% loadings did not result in significant improvement in UV resistance. Visual inspection revealed a yellowing level comparable to that of the reference material, and tensile stress and breaking force values after 240 h were similar to those of the unmodified samples.
- Addition of TiO_2 – proved clearly more effective. Particularly at 3% loading, it reduced the rate of yellowing and improved the retention of mechanical strength. After 240 h of UV exposure, TiO_2-modified samples exhibited the highest resistance among all analysed variants.
- Mechanisms of action – the findings indicate that TiO_2 primarily acts as an optical UV shield (absorption and scattering), whereas Al_2O_3 contributes mainly via barrier

and mechanical reinforcement effects, which under the studied conditions did not provide sufficient protection against photodegradation.

- Practical significance – the best protective effects were achieved with 3% TiO_2. This outcome is consistent with literature reports, which highlight low to moderate nanofiller contents (0.5–3 wt.%) as most beneficial for retaining the properties of polymers and composites exposed to UV conditions.
- Directions for further research – it is necessary to extend the analyses to longer exposure times and real environmental conditions (variable thermal and humidity cycles), as well as to employ additional analytical methods (e.g., FTIR, SEM, DMA) to more precisely identify degradation mechanisms and verify the role of metal oxides in stabilizing epoxy systems.

References

Abdeen, D.H., El Hachach, M., Koc, M., Atieh, M.A.: A Review on the Corrosion Behaviour of Nanocoatings on Metallic Substrates. Materials **12**, 210 (2019). https://doi.org/10.3390/ma12020210

Affrossman, S., McKee, A., Pethrick, R.A.: Effect of surface treatments of titanium dioxide pigments on the cure of polyester/triglycidyl isocyanurate powder coatings. J. Coat. Technol. Res. **12**, 1053–1064 (2015). https://doi.org/10.1007/s11998-015-9697-9

Amorim, F.C., Reis, J.M.L., Souza, J.F.B., Da Costa Mattos, H.S.: Investigation of UV exposure in adhesively bonded single lap joints. Appl. Adhes. Sci. **6**, 2 (2018). https://doi.org/10.1186/s40563-018-0103-6

Andrady, A.L., et al.: Oxidation and fragmentation of plastics in a changing environment; from UV-radiation to biological degradation. Sci. Total Environ. **851**, 158022 (2022). https://doi.org/10.1016/j.scitotenv.2022.158022

Andrady, A.L., et al.: Effects of UV radiation on natural and synthetic materials. Photochem. Photobiol. Sci. **22**, 1177–1202 (2023). https://doi.org/10.1007/s43630-023-00377-6

Bazrgari, D., Moztarzadeh, F., Sabbagh-Alvani, A.A., Rasoulianboroujeni, M., Tahriri, M., Tayebi, L.: Mechanical properties and tribological performance of epoxy/Al2O3 nanocomposite. Ceram. Int. **44**, 1220–1224 (2018). https://doi.org/10.1016/j.ceramint.2017.10.068

Chamas, A., et al.: Degradation Rates of Plastics in the Environment. ACS Sustain. Chem. Eng. **8**, 3494–3511 (2020). https://doi.org/10.1021/acssuschemeng.9b06635

Chen, X., Mao, S.S.: Titanium Dioxide Nanomaterials: Synthesis, Properties, Modifications, and Applications. Chem. Rev. **107**, 2891–2959 (2007). https://doi.org/10.1021/cr0500535

Chen, Y., Liu, R., Luo, J.: Improvement of anti-aging property of UV-curable coatings with silica-coated TiO2. Prog. Org. Coat. **179**, 107479 (2023). https://doi.org/10.1016/j.porgcoat.2023.107479

Ci, S., Wang, B., Di, C., Wang, M., Zhu, B., Qiao, K.: Effect of Ultraviolet Aging on Properties of Epoxy Resin and Its Pultruded Fiber-Reinforced Composite. Polymers **17**, 294 (2025). https://doi.org/10.3390/polym17030294

Dhoke, S.K., Khanna, A.S.: Effect of nano-Fe2O3 particles on the corrosion behavior of alkyd based waterborne coatings. Corros. Sci. **51**, 6–20 (2009). https://doi.org/10.1016/j.corsci.2008.09.028

Egerton, T.: The Influence of Surface Alumina and Silica on the Photocatalytic Degradation of Organic Pollutants. Catalysts **3**, 338–362 (2013). https://doi.org/10.3390/catal3010338

Fernández-Álvarez, M., Velasco, F., Bautista, A.: Performance of ultraviolet exposed epoxy powder coatings functionalized with silica by hot mixing. J. Mater. Res. Technol. **10**, 1042–1057 (2021). https://doi.org/10.1016/j.jmrt.2020.12.094

Goyat, M.S., Rana, S., Halder, S., Ghosh, P.K.: Facile fabrication of epoxy-TiO2 nanocomposites: A critical analysis of TiO2 impact on mechanical properties and toughening mechanisms. Ultrason. Sonochem. **40**, 861–873 (2018). https://doi.org/10.1016/j.ultsonch.2017.07.040

Gültekin, K., Uğuz, G., Topcu, Y., Özel, A.: Structural, thermal, and mechanical properties of silanized boron carbide doped epoxy nanocomposites. J. Appl. Polymer Sci. **138**, 51244 (2021). https://doi.org/10.1002/app.51244

Kim, N., et al.: Surface Coating of Titanium Dioxide Nanoparticles with a Polymerizable Chelating Agent and Its Physicochemical Property. ACS Omega **8**, 18743–18750 (2023). https://doi.org/10.1021/acsomega.3c00734

Korkmaz, Y., Gültekin, K.: Effect of UV irradiation on epoxy adhesives and adhesively bonded joints reinforced with BN and B4C nanoparticles. Polym. Degrad. Stab. **202**, 110004 (2022). https://doi.org/10.1016/j.polymdegradstab.2022.110004

Martins, T.D., Ribeiro, T., Farinha, J.P.S.: Overview of Silica-Polymer Nanostructures for Waterborne High-Performance Coatings. Polymers **13**, 1003 (2021). https://doi.org/10.3390/polym13071003

Miturska, I., Rudawska, A., Müller, M., Valášek, P.: The Influence of Modification with Natural Fillers on the Mechanical Properties of Epoxy Adhesive Compositions after Storage Time. Materials **13**, 291 (2020). https://doi.org/10.3390/ma13020291

Nakata, K., Fujishima, A.: TiO2 photocatalysis: Design and applications. J. Photochem. Photobiol., C **13**, 169–189 (2012). https://doi.org/10.1016/j.jphotochemrev.2012.06.001

Nguyen, T.V., Nguyen, T.A., Nguyen, T.H.: The Synergistic Effects of Sio2 Nanoparticles and Organic Photostabilizers for Enhanced Weathering Resistance of Acrylic Polyurethane Coating. J. Compos Sci. **4**, 23 (2020). https://doi.org/10.3390/jcs4010023

Nikafshar, S., Zabihi, O., Ahmadi, M., Mirmohseni, A., Taseidifar, M., Naebe, M.: The Effects of UV Light on the Chemical and Mechanical Properties of a Transparent Epoxy-Diamine System in the Presence of an Organic UV Absorber. Materials **10**, 180 (2017). https://doi.org/10.3390/ma10020180

Pan, Z., Liu, Y., Wang, F., Lu, G., Yang, F., Cheng, F.: Al_2 O_3 Dispersion-Induced Micropapillae in an Epoxy Composite Coating and Implications in Thermal Conductivity. ACS Omega **6**, 17870–17879 (2021). https://doi.org/10.1021/acsomega.1c01282

Petković, G., Pasanec Preprotić, S., Vukoje, M., Bolanča Mirković, I.: Impact of SiO2 and TiO2 Nanoparticles on the Elasticity and Aging Resistance of Polyvinyl Acetate (PVAc) Adhesive. Materials **17**, 5957 (2024). https://doi.org/10.3390/ma17235957

Petković, G., Vukoje, M., Bota, J., Pasanec Preprotić, S.: Enhancement of Polyvinyl Acetate (PVAc) Adhesion Performance by SiO2 and TiO2 Nanoparticles. Coatings **9**, 707 (2019). https://doi.org/10.3390/coatings9110707

Singla, P., Mehta, R., Upadhyay, S.N.: Microwave assisted in situ ring-opening polymerization of polylactide/clay nanocomposites: Effect of clay loading. Appl. Clay Sci. **95**, 67–73 (2014). https://doi.org/10.1016/j.clay.2014.03.012

Şomoghi, R., et al.: The Impact of ZnO Nanofillers on the Mechanical and Anti-Corrosion Performances of Epoxy Composites. Polymers **16**, 2054 (2024). https://doi.org/10.3390/polym16142054

Wei, Y., Jin, X., Luo, Q., Li, Q., Sun, G.: Adhesively bonded joints – A review on design, manufacturing, experiments, modeling and challenges. Compos. B Eng. **276**, 111225 (2024). https://doi.org/10.1016/j.compositesb.2024.111225

Wojciechowski, K., Zukowska, G.Z., Korczagin, I., Malanowski, P.: Effect of TiO2 on UV stability of polymeric binder films used in waterborne facade paints. Prog. Org. Coat. **85**, 123–130 (2015). https://doi.org/10.1016/j.porgcoat.2015.04.002

Xia, R., Xi, J., Zhang, Z., He, Y., Yu, Z.: Curing Behavior of UV-Initiated Surface-Modified Nano-TiO2/Epoxy Resin Prepolymers and the Properties of Cured Composites. Polymers **15**, 1756 (2023). https://doi.org/10.3390/polym15071756

Yousif, E., Haddad, R.: Photodegradation and photostabilization of polymers, especially polystyrene: review. Springerplus **2**, 398 (2013). https://doi.org/10.1186/2193-1801-2-398

Zeljko, M., Ocelić Bulatović, V., Špada, V., Blagojević, S.L.: Environmentally Friendly UV-Protective Polyacrylate/TiO2 Nanocoatings. Polymers **13**, 2609 (2021). https://doi.org/10.3390/polym13162609

Zhang, Y., et al.: High-Performance Al2O3/Epoxy Resin Composites for Insulating Pull Rods of Direct Current High-Speed Switches. Energies **18**, 4256 (2025). https://doi.org/10.3390/en18164256

Relationship Between Activation Energy and Reaction Enthalpy in Urea Formaldehyde Resin

Latifeh Nasseri[1,2(✉)], Roland Mitter[3], Johann Moser[4], Andreas Kandelbauer[5], Johannes Konnerth[1], and Hendrikus WG. Van Herwijnen[1,2]

[1] Department of Natural Sciences and Sustainable Resources, Institute of Wood Technology and Renewable Materials, BOKU University, Konrad Lorenz Strasse 24, 3430 Tulln, Austria
lnasseri@groupwise.boku.ac.at

[2] Wood K Plus - Competence Centre for Wood Composites and Wood Chemistry, Kompetenzzentrum Holz GmbH, Altenberger Straße 69, 4040 Linz, Austria

[3] Fritz Egger, GmbH and Co OG, Tiroler Straße 16, 3105 Unterradlberg, Austria

[4] KremsChem Austria GmbH, Hafenstrasse 77, 3500 Krems, Austria

[5] Center for Process Analysis and Technology (PA&T), School of Life Sciences, Reutlingen University, Alteburgstrasse 150, 72762 Reutlingen, Germany

Abstract. The curing process of urea-formaldehyde (UF) resins consists of parallel, consecutive, and competitive reactions, which complicates defining a single activation energy or reaction type. Differential Scanning Calorimetry (DSC) was used to measure reaction enthalpy in joules per gram (J g^{-1}). By conducting DSC experiments at three heating rates, the activation energy, expressed in joules per mole (J mol^{-1}), was determined as a function of conversion degree using the model-free Friedman method. For such multistep reactions, the calculated activation energy represents an average of several reaction steps. The complex composition of UF resin samples further complicates the quantification of "one mole" of adhesive, raising the question of how many moles correspond to the measured enthalpy. To address this, dimethylolurea (DMU) was selected as a simplified model compound. The methylol group, a reactive unit in DMU with known enthalpy and defined molar mass, allowed the establishment of a relationship between activation energy and enthalpy in UF reactions. Since methylol groups are key participants in UF resin curing, they served as representative units for analysis. The results showed that activation energy is considerably higher than reaction enthalpy, emphasizing the dominant role of kinetic barriers in UF curing. These findings provide new insights into UF resin reaction kinetics.

Keywords: enthalpy · activation energy · curing kinetics · urea-formaldehyde resin · methylol group

1 Introduction

The chemistry of urea-formaldehyde (UF) resin synthesis can be divided into two main reaction mechanisms: methylolation and condensation reactions. During the methylolation stage, urea and formaldehyde undergo addition reactions, leading to the formation

L. F. M. da Silva and R. D. Adams (Eds.): AB 2025, PEM, pp. 19–32, 2026.
https://doi.org/10.1007/978-3-032-19022-2_2

of mono-, di-, and tri-methylol ureas. The methylolation stage can occur under either acidic or basic conditions. Each type of methylolation is characterized by its own rate constant, with distinct values for the forward and reverse reactions. The reversible nature of methylolation step lowers hydrolysis resistance, enabling moisture-induced formaldehyde release from weakly bonded groups. The second stage, condensation, involves the reaction of the methylols with each other or with urea to form linear and branched polymeric material [1]. Different formaldehyde to urea (F/U) molar ratios and curing conditions (e.g., temperature and pH) influence the chemical linkages and branching of the polymer network [2]: at lower temperatures and mildly acidic pH, methylene ether bridges (—CH_2—O—CH_2—) are preferentially formed, whereas higher temperatures and more strongly acidic environments promote the formation of the more stable methylene (—CH_2—) bridges [1, 3]. In addition, the methylene ether bridges can be converted into methylene bridges and release of formaldehyde.

The structure of UF resin and the overall curing process is quite complex [4, 5], involving many parallel, consecutive, and competitive reactions [6]. Consequently, the actual sequence and individual steps of this reaction are complex, making it impossible to define a single activation energy value for such complex multi-step reactions.

There are different approaches for evaluating the thermal cure kinetics: model-based and model-free [7]. The model-based method is used when each reaction step can be defined and has its kinetic equation and kinetic triplet containing reaction type, activation energy, and pre-exponential factor. On the other hand, the model-free approach is suitable for more complex reactions where no initial kinetic information is available to suggest the correct model with accurate reaction types and parameters. In this approach, kinetic analysis is possible without defining the reaction type, as long as data are compared at identical levels of conversion [8]. The model-free isoconversional Friedman method determines the dependency of the activation energy on the degree of conversion [7]. In our previous publication [9], we aimed to calculate the activation energy of formaldehyde-based resins with different curing techniques, such as Differential Scanning Calorimetry (DSC), rheometry, and the Automated Bonding Evaluation System (ABES), using the model-free Friedman method [10]. Since DSC focuses solely on the chemical aspect of curing, this technique is employed in the present study. During dynamic DSC runs, the sample undergoes controlled heating at a constant rate over a specified temperature range. DSC enables the measurement of heat flow from a sample as a function of time or temperature. The mass of the sample in grams is the input value in DSC. The area of the peak of the heat flow curve represents the reaction enthalpy (ΔH). Changes in reaction enthalpy arise from exothermic and endothermic events in the sample and are given in joules per gram of the sample (J g^{-1}). Polycondensation reactions, in particular, can be either exothermic or endothermic in nature. The chemical cure conversion (α_{DSC}) is the ratio of the heat released at a given time t (ΔH_t) in the sample from the onset of polymerization or curing at time t_i to the total heat for the complete conversion (ΔH_{total}) at time t_f:

$$\alpha(\mathrm{t})_{\mathrm{DSC}} = \frac{\int_{\mathrm{t_i}}^{\mathrm{t}} \left(\frac{\mathrm{dH}}{\mathrm{dt}}\right)\mathrm{dt}}{\int_{\mathrm{t_i}}^{\mathrm{t_f}} \left(\frac{\mathrm{dH}}{\mathrm{dt}}\right)\mathrm{dt}} = \frac{\Delta H_{\mathrm{t}}}{\Delta H_{\mathrm{total}}} \tag{1}$$

The detailed calculation of the activation energy according to the model-free Friedman method for non-isothermal conditions with a heating rate of $\beta(= dT/dt)$ is explained elsewhere [9]. In model-free kinetics, the activation energy is calculated in joules per mole at a specific conversion degree and defines the energy barrier that the reaction steps need to overcome to progress further.

In the kinetics of thermosetting polymers, both the activation energy and the reaction enthalpy are essential parameters, reflecting kinetic and thermodynamic aspects of the cure, respectively. Mashouf Roudsari et al. [11] demonstrated that altering epoxy curing chemistry simultaneously influenced both ΔH and activation energy, indicating changes in network thermodynamics and cure kinetics. Zhang et al. [12] studied UF and melamine-urea-formaldehyde (MUF) resins and found that incorporating melamine into the resin reduced the total reaction enthalpy while simultaneously modifying the cure kinetics. Also when UF resins are modified with just a little melamine (mUF resins) a lower reaction enthalpy and lower activation energy is found. These highlight that activation energy and enthalpy provide complementary but distinct insights, both of which are necessary for a comprehensive understanding of resin curing.

In simple reactions, where the elementary steps and molar masses are well-defined, the reaction enthalpy of the forward and backward reactions can be determined, and the corresponding activation energies and pre-exponential factors can be calculated. This enables quantitative relationships between activation energy and reaction enthalpy to be established, as shown in [13, 14]. In contrast, the diverse structural units in UF resins, including methylene and ether bridges, methylol and amide groups, and cyclic derivatives such as uron rings, combined with the multitude of possible reaction pathways, makes their curing study highly complex [6]. For such complex systems, the calculated activation energy in joules per mole represents an average over multiple steps.

When analyzing the curing kinetics of UF resin, each gram of the sample consists of a complex adhesive mixture, making it challenging to define and quantify "one mole" of the adhesive system. Additionally, varying formaldehyde-to-urea molar ratios and differences in molar mass across resin systems complicate reactivity comparisons [15]. To the best of our knowledge, what the activation energy in joules per mole signifies in relation to the amount of UF resin used has not been previously reported in the literature. The key question is: how many moles correspond to the enthalpy of this sample? As there is no information about the moles of the sample (UF resin), the conversion between joules per gram and joules per mole is not possible. However, if the enthalpy of an intermediate molecule with a defined molar mass, actively involved in the UF resin reactions, could be identified, this relation would become accessible. In this context, the methylol group is of particular interest, as its content is a key determinant of UF resin reactivity [21, 22]. Chow and Steiner [23] identified the methylol group as the reactive site by applying bromination in CCl_4 followed by X-ray analysis. In addition, high-resolution nuclear magnetic resonance (NMR) has been applied to UF resins to monitor their structural features and trace their kinetic behavior, further supporting the central role of methylol groups in curing reactions [16]. Therefore, in this study, the methylol group was selected as the focal point for kinetic analysis, serving as a representative unit to explore the relationship between enthalpy and activation energy during UF resin curing. Establishing this relation offers a pathway to a more precise understanding of UF resin reaction kinetics.

2 Experimental

2.1 Chemicals

Dimethylolurea (DMU) powder (purity (TLC) $\geq$ 95%), urea, and ammonium nitrate were purchased from Sigma-Aldrich (Merck Supelco). The physicochemical properties of DMU are given in Table 1. An industrial urea formaldehyde resin (UF2), typically used for particleboard core layer fulfilling the E1 (EN 13986) (39) formaldehyde emission requirements, was supplied by KremsChem Austria GmbH, Krems, Austria. The molar ratio and solid contents of UF2 are listed in Table 2.

Table 1. Physicochemical properties of dimethylolurea (DMU)

DMU (1,3-Bis(hydroxymethyl)urea)	
chemical formula	$C_3H_8N_2O_3$
molecular weight	120.1 g mol^{-1}
melting point	118.6 °C
decomposition temperature	$\geq$ 118.6 °C

2.2 Sample Preparation

Samples with different compositions for DSC analysis are listed in Table 2. Aqueous ammonium nitrate solution (15% w/w) was added as a hardener in a weight ratio of 10% w/w. DMU and urea were mixed in the molar ratio of 1:1 and the target solid content of all samples was 66 wt%, equal to the solid content of the tested liquid UF resin (UF2), to ensure an essential comparison. Experiments were conducted with at least three repetitions to ensure consistency and adequacy of the conclusions.

Table 2. Samples used for DSC experiments. Star (*) presents a specific part of the composition

Sample	Composition	Hardener: *(liquid: liquid)	Molar ratio
Sample 1	DMU + H_2O	no hardener	F/U = 2
Sample 2	* (DMU + H_2O) + NH_4NO_3(15%)	1:10	F/U = 2
Sample 3	DMU + Urea + H_2O	no hardener	F/U = 1
Sample 4	* (DMU + Urea + H_2O) + NH_4NO_3(15%)	1:10	F/U = 1

(continued)

Table 2. (*continued*)

Sample	Composition	Hardener: *(liquid: liquid)	Molar ratio
Sample 5	* (DMU + Urea + H_2O) + NH_4NO_3(15%)	1:20	F/U = 1
Sample 6	* (UF2 resin) + NH_4NO_3(15%)	1:10	F/U = 1

2.3 DSC Measurement

The thermal behavior of the samples was investigated using differential scanning calorimetry (DSC) (Polyma 214, Netzsch-Gerätebau GmbH, Germany). Approximately 11–12 mg of each sample was weighed into a sealed high-pressure stainless steel (27 μl) crucible and analyzed by DSC from -70 °C to 230 °C at heating rates of 1, 2, and 5 °C min^{-1}. Three DSC runs were performed for each heating rate.

2.4 TGA-IR Measurement

TGA-IR measurements were carried out using a TA Instruments TGA Q5000IR coupled with a Thermo Electron Corporation Nicolet 380 FT-IR spectrometer. Samples (~5 mg) were placed in a high-temperature platinum crucible and heated from room temperature to 600 °C at a rate of 5 °C min^{-1} under a nitrogen atmosphere (25 mL min^{-1}). The experiments were performed in duplicate to ensure reproducibility of the results.

2.5 Analysis of Thermochemical Data

Enthalpy analyses were performed using Netzsch Proteus (Netzsch-Gerätebau) and OriginPro 2025 (OriginLab Corporation, Northampton, Massachusetts, USA). Kinetic analysis according to the model-free Friedman method was performed using NETZSCH Kinetics Neo (v. 2.6.0.1) software. To reduce the impact of high relative errors at extreme conversion levels, kinetic analysis was limited to a conversion range of 0.10 - 0.90 [7].

3 Results and Discussion

As schematically illustrated in Fig. 1, in the methylolation stage of UF resin synthesis, formaldehyde reacts rapidly with urea. N', N''-DMU is an intermediate product formed during the methylolation stage, characterized by a defined molar mass. It is incorporated in the polycondensation reaction of UF resin and, as such, serves as an ideal model substance for our study. UF resins comprise a wide range of structural elements, including methylene and ether bridges, methylol and amide groups, as well as cyclic derivatives such as uron rings. This structural heterogeneity, together with the numerous parallel and competing reactions, accounts for the difficulty in interpreting their curing behavior [6].

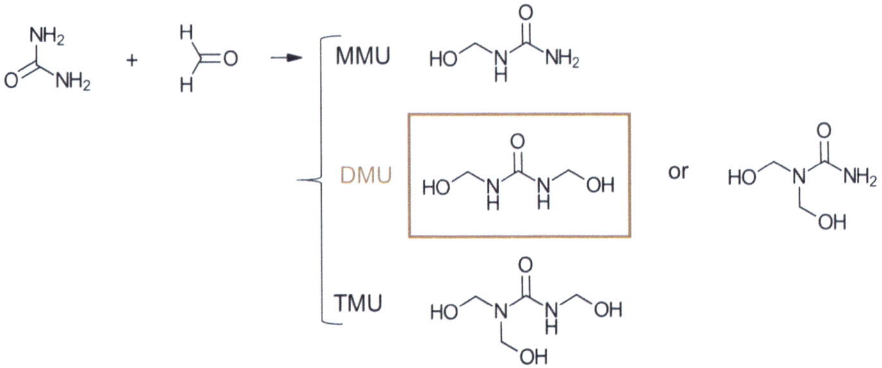

Fig. 1. Example of the addition of formaldehyde to urea to mono-, di-, and tri-methylolurea

Figure 2 presents the DSC thermogram of the tested UF resin (UF2) cured dynamically at 5 °C min^{-1}. The total heat for complete conversion (ΔH_{total}), calculated as the area under the DSC curve, is 114 J g^{-1}. The resulting DSC thermogram is comprised of the combined signals of all occurring reactions. By performing dynamic runs with at least three heating rates, the average activation energy can be calculated in J mol^{-1}.

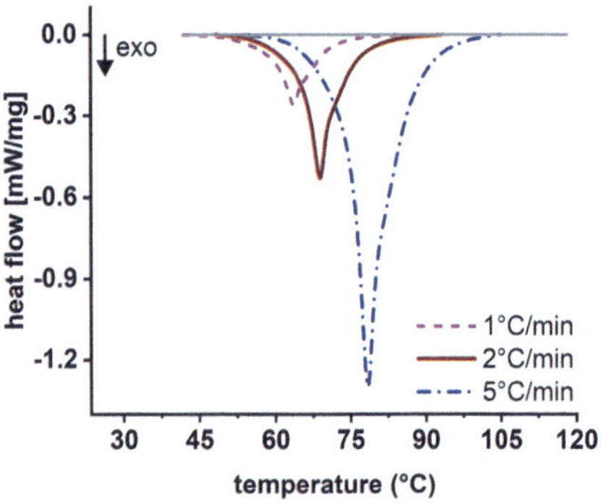

Fig. 2. Normalized heat flow of the tested UF resin (UF2, sample 6) measured dynamically using DSC at three different heating rates

As schematically illustrated in Fig. 3, at the initial stage of condensation, DMU can theoretically react with itself (I), with urea (II), or decompose into monomethylolurea and formaldehyde (III). The condensation reactions of DMU lead to form two different types of linkages, methylene and methylene ether bridges and to produce low molecular weight polymers.

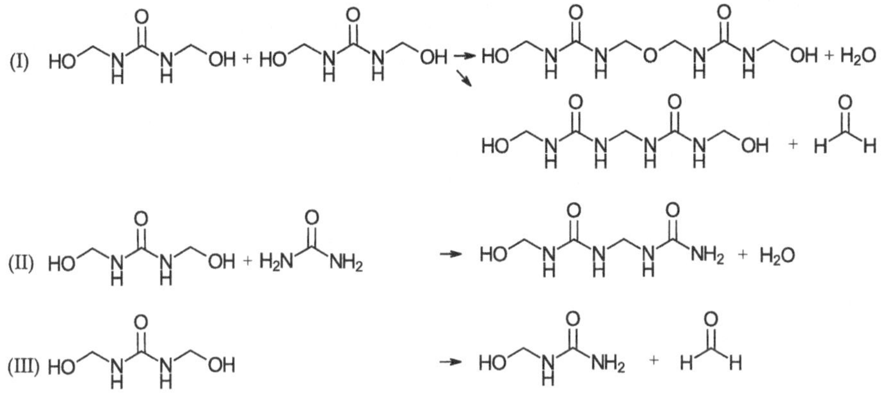

Fig. 3. Schematic representation of potential condensation reactions involving DMU

In the 1950s de Jong and de Jonge [17–20] extensively studied the kinetics and mechanisms of the condensation reactions between urea and formaldehyde. De Jong and de Jonge [17] demonstrated through their calculations that the dissociation of DMU into MMU and formaldehyde (III) is negligible within the pH range of 3.6 to 4.5. In contrast, Nair et al. [21] showed that DMU dissociates to monomethylol urea and formaldehyde in aqueous solution, in acid, neutral and alkaline conditions. In this study, ammonium nitrate was introduced as a catalyst. In Fig. 4, the second endothermic peak observed in samples 1, 2, and 4 is probably associated with the formaldehyde release due to the reaction of DMU with itself (I), or DMU decomposition (III). To verify this assumption, TGA-IR measurements were conducted on a pure DMU sample (powder).

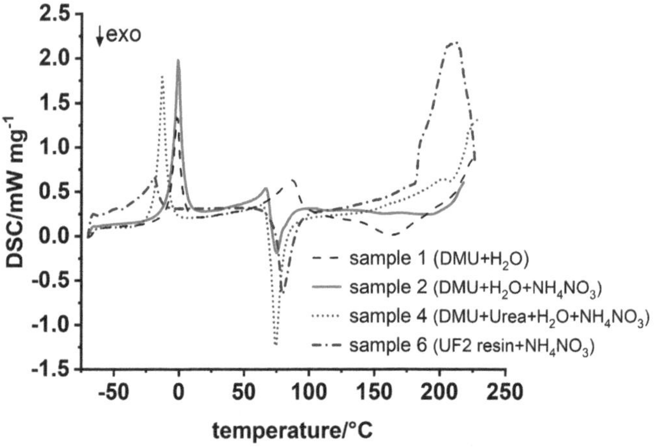

Fig. 4. DSC thermogram of sample 1 (dashed line), sample 2 (solid line), sample 4 (dotted line), and sample 6 (dashdotted line). Dynamic DSC runs were performed at a heating rate of 5 °C min^{-1}

Figure 5 presents the results of TGA-FTIR measurements. The assignment of characteristic peaks was done based on relevant literature [22–24] (Table 3). In the first decomposition stage, a pronounced C = O band at ~ 1745 cm^{-1} together with the characteristic CH stretching doublet at ~ 2840 and ~ 2780 cm^{-1} is observed. This is evidence for the early release of formaldehyde.

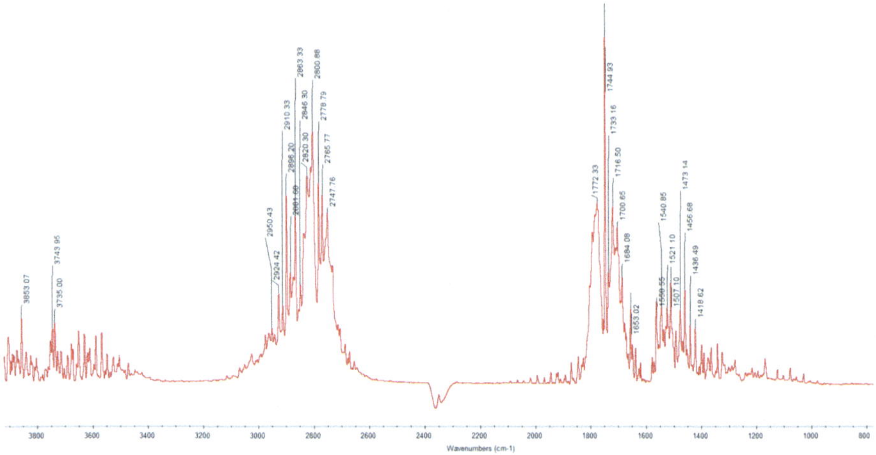

Fig. 5. IR spectrum of the first decomposition stage of DMU powder in the TG experiment

Table 3. Comparison of formaldehyde IR bands from literature with those observed in the first decomposition stage (Fig. 5)

Literature data [22–24] wavenumber, cm^{-1}	Observed wavenumber, cm^{-1}	Functional group
3400	~ 3350	O–H stretch (possible overlap with H_2O)
2822	~ 2840	CH stretch (aldehyde, diagnostic)
2724	~ 2780	CH stretch (aldehyde, diagnostic)
1722	~ 1745	C = O stretch (formaldehyde)
1390	~ 1440–1500	C–H bend

De Jong and de Jonge [17] also claimed that the self-condensation of DMU (I) was very slow, if occurring at all. Nair et al. [21] reported that the self-condensation of DMU was slower than the reaction of DMU with urea (II). However, the DSC thermograms show an exothermic peak, likely associated with the formation of methylene or ether bridges, confirming the reaction between DMU molecules (I). Siimer et al. [25] claimed that in curing without a catalyst, in mildly alkaline medium (resin pH 8.2), polycondensation takes place preferably over the formation of methylene linkages between linear resin fragments. Also, the endothermic decomposition of DMU yields MMU (III), which then subsequently react exothermically with urea (II). Furthermore, the formed MMU has a free amine group that can react exothermically with DMU (not depicted in Fig. 3).

Siimer et al. [25] claimed also that in the presence of an acid catalyst, the first exotherm arises from a polycondensation reaction of primary amino groups of unreacted urea with methylol groups of resin. Szesztay et al. [26] analyzed DSC thermograms of three industrial UF resins, noting that the exothermic peaks before 100 °C were due to polycondensation reactions. They claimed that these peaks, consistent across samples at 80 - 85 °C, correspond to crosslinking reactions with similar reaction heats. However, their results showed that the endothermic reactions from 100 - 150 °C varied, likely due to the evaporation of initial and reaction water, formaldehyde release, and decomposition of methylene ether linkage [26].

Due to the simultaneous occurrence of multiple reactions during UF resin curing, individual pathways cannot be clearly defined. A simpler model substance is therefore required to simulate specific aspects of the curing process. While the exothermic peak onsets for sample 4 containing DMU and sample 6 (UF2 resin) are similar, sample 4 as the simpler model substance, exhibits a more distinct and well-defined composition. Moreover, the exothermic peak observed in sample 4 is a subset of the broader exothermic peak seen in sample 6. Therefore, to calculate the relation between J g^{-1} and J mol^{-1}, sample 4 was used. Consequently, the reaction kinetics is modeled by performing dynamic DSC measurements using this sample (see Fig. 6 (left)). Figure 6 (right) presents the plot of activation energy as a function of the extent of conversion (α_{DSC}). The average activation energy increases as the reaction proceeds. For sample 4 the mean value of the enthalpy for all heating rates is $\overline{\Delta H} = 131.7$ J g^{-1}, while the mean value of the resulting activation energy is $\overline{E_\alpha} = 93$ kJ mol^{-1}.

Nair et al. [21] mentioned that the rate of addition of formaldehyde to MMU is lower compared to its reaction with free urea. This reduced reactivity is primarily due to a statistical factor: while urea has two reactive positions, monomethylol urea (MMU) has only one. Furthermore, for forming higher homologues in UF reaction, the condensation of methylol groups with free urea appears to be the dominant route [21].

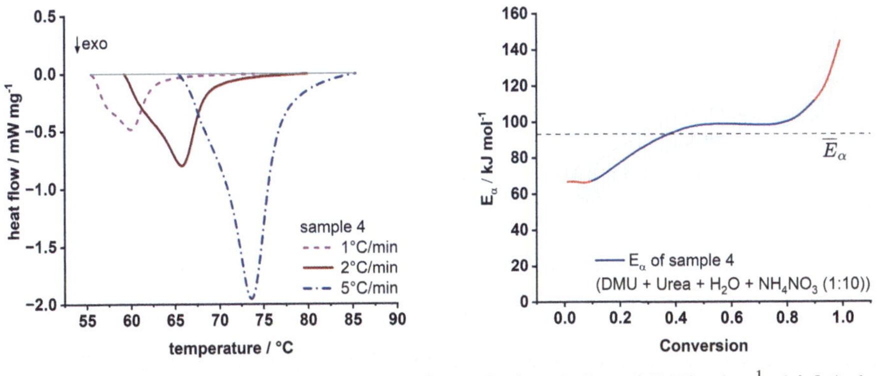

Fig. 6. **(left)** Dynamic DSC measurements of sample 4 at 1, 2, and 5 °C min^{-1}, **(right)** the activation energy dependence on the degree of conversion for sample 4

Assuming the methylol group as the reactive group [27, 28] with known enthalpy allows linking activation energy to enthalpy in the initial stages of the UF reactions. Therefore, it is necessary to calculate the number of moles of methylol groups present in the sample. One gram of sample 4 (DMU + Urea + H_2O + NH_4NO_3 (15%)), contains 0.4 g (wt%) DMU, which corresponds to 3.3 mmol of DMU. As one mole of DMU comprises two moles of methylol groups, sample 4 contains 6.7 mmol of methylol groups per gram. Consequently, the enthalpy of the reactions of methylol groups can be calculated as follows:

$$\Delta H(CH_2OH) = \frac{\overline{\Delta H}}{n_{methylolgroup}} = \frac{131.7Jg^{-1}}{6.7E^{-03}\ \text{mol}\ g^{-1}} = 19.8kJmol^{-1}$$

Accordingly, the relation between the activation energy and the resulting reaction enthalpy can be calculated as follows:

$$\frac{\overline{E_\alpha}}{\Delta H(CH_2OH)} = \frac{93.0kJmol^{-1}}{19.8kJmol^{-1}} = 4.7$$

A similar relationship was calculated for samples 5 and 3, as summarized in Table 4. This shows that, in all examples, the activation energy is clearly higher than the resulting enthalpy ($\Delta H(CH_2OH)$). This confirms that the energy barrier to initiate the reaction remains the controlling factor despite the thermodynamic favorability of the process.

Figure 7 shows the change in (average) activation energy over time. The activation energy profile of UF resin (sample 6) closely follows those of the other samples containing DMU, further supporting the validity of using DMU as the model substance in this study. When no hardener is used (sample 3), the average activation energy is the highest. This is particularly evident at the beginning of the reaction, where the activation energy is significantly elevated (Fig. 7). This suggests that the acid catalyst primarily activates the methylol group, with limited influence on subsequent reaction stages. Upon the addition of hardener at a 1:20 ratio (sample 5), the catalyst effectively lowers the activation energy. Furthermore, the overall reaction enthalpy increases, indicating that additional reactions, such as the decomposition of ether bridges with subsequent reactions, are taking place. Interestingly, increasing the hardener concentration to a 10:1 ratio (sample 4) does not lead to a substantial further reduction in activation energy, suggesting that excess hardener is unnecessary. The latent hardener not only releases nitric acid but also produces ammonia. This ammonia captures the highly reactive formaldehyde formed during ether bridge decomposition, making it unavailable for subsequent reactions to new methylol groups, which leads to a decrease in overall reaction enthalpy and an increase in activation energy. Notably, in the early stages of the reaction, the differences in activation energy between samples 4 and 5 are minimal.

Table 4. Activation energy, enthalpy and the ratio between them for the methylol group (CH_2OH) in samples 3, 4, and 5

Sample	$\overline{E_\alpha}$/kJ mol^{-1}	$\overline{\Delta H}$/J g^{-1}	$\Delta H(CH_2OH)$/kJ mol^{-1}	$\frac{\overline{E_\alpha}}{\Delta H(CH_2OH)}$
Sample 3 (no hardener)	98.4	127 ± 12	17.3	5.7
Sample 4 (hardener 1:10)	93.0	132 ± 14	19.8	4.7
Sample 5 (hardener 1:20)	85.5	149 ± 10	21.3	4.0

De Jong and de Jonge [17] calculated the activation energy for the addition reaction between DMU and urea at pH of 3.8 as 15 kcal mol^{-1} or 62.8 kJ mol^{-1}, based on titration data. This comparatively low value can likely be attributed to their use of an acidic acetate buffer solution, which is presumably more catalytically active than the latent catalyst employed in the present study.

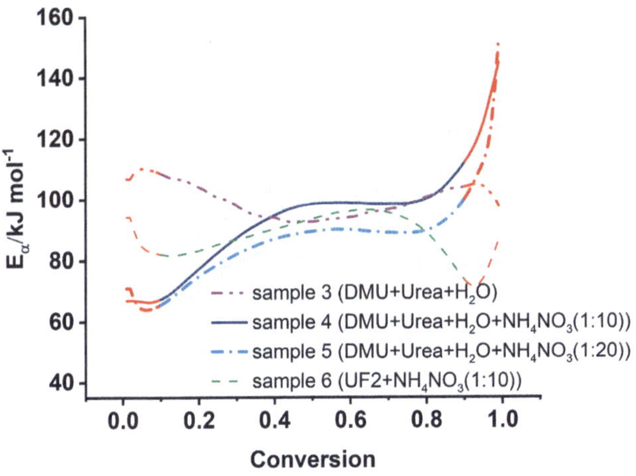

Fig. 7. The dependence of activation energy on the degree of conversion for samples 3, 4, 5, and 6

De Jong and de Jonge [20] also calculated the activation energy for the preceding addition reaction between urea and formaldehyde and the dissociation reaction as 54.4 kJ mol^{-1} and 79.5 kJ mol^{-1}, respectively. As the former has a lower activation energy, the addition reaction is more favorable than the decomposition. Subtracting these activation energies yields a reaction enthalpy of 25.1 kJ mol^{-1}, which is close to the reactions occurring of this study. Figure 8 provides a schematic representation of the relationship between activation energy and the enthalpy for sample 4.

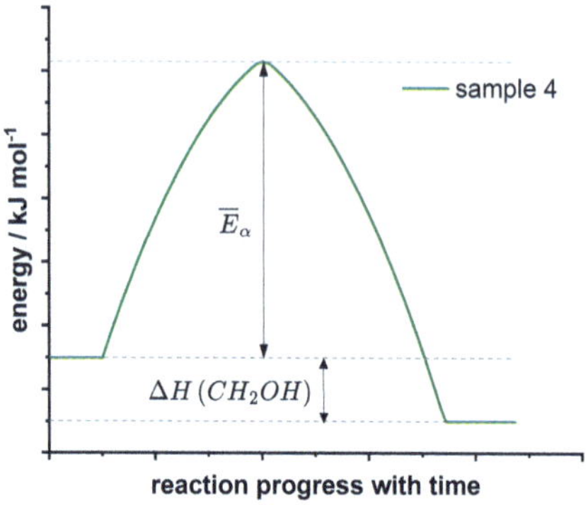

Fig. 8. Schematic representation of the relationship between the enthalpy and activation energy calculated for sample 4

4 Conclusion

Due to the complex structure of UF resins, quantifying the exact number of moles of reactants converted into products is not feasible, as the precise composition of the resin remains unclear. Using DMU as a model substance having two methylol groups, with a defined molar mass and reaction pathway for kinetic analysis provides at least a minimal insight into the relationship between enthalpy and activation energy of UF resin curing. The results show that the activation energy is significantly higher than the reaction enthalpy. This underlines that kinetic barriers, rather than thermodynamic limitations, are the primary factors controlling the initial phase of the curing process of UF resins.

Acknowledgments. The authors gratefully acknowledge Sandra Jury for performing the TGA-IR measurements. This research was conducted within the framework of a project supported by the Austrian Research Promotion Agency (FFG) under the COMET program (Project No. 892416).

References

1. Dunky, M.: Urea–formaldehyde (UF) adhesive resins for wood. Int. J. Adhesion and Adhesives **18**(2), 95–107 (1998). https://doi.org/10.1016/S0143-7496(97)00054-7
2. Dunky, M.: Technology of Adhesives and Wood-based Panels. Wiley Online Library (2025)
3. Kibrik, É.J., Steinhof, O., Scherr, G., Thiel, W.R., Hasse, H.: On-Line NMR Spectroscopic Reaction Kinetic Study of Urea–Formaldehyde Resin Synthesis. Indus. Eng. Chem. Res. **53**(32), 12602–13 (2014). https://doi.org/10.1021/ie5001746
4. Conner, A.H.: Urea-formaldehyde adhesive resins. Polymeric Materials Encyclopedia **11**(8) (1996)
5. Jada, S.S.: The structure of urea—formaldehyde resins. J. Appl. Polymer Sci. **35**(6), 1573–92 (1988). https://doi.org/10.1002/app.1988.070350614
6. Carvalho, L.M.H., Costa, M.R.P.F.N., Costa, C.A.V.: A very simple empirical kinetic model of the acid-catalyzed cure of urea–formaldehyde resins. J. Appl. Polymer Sci. **102**(6), 5977–87 (2006). https://doi.org/10.1002/app.25174

7. Vyazovkin, S., et al.: ICTAC Kinetics Committee recommendations for performing kinetic computations on thermal analysis data. Thermochimica Acta. **520**(1), 1–19 (2011). https://doi.org/10.1016/j.tca.2011.03.034
8. Vyazovkin, S., et al.: ICTAC Kinetics Committee recommendations for analysis of multi-step kinetics. Thermochimica Acta **689**, 178597 (2020). https://doi.org/10.1016/j.tca.2020.178597
9. Nasseri, L., et al.: Comparison between cure kinetics by means of dynamic rheology and DSC of formaldehyde-based wood adhesives. The J. Adhesion. 1–22 (2024). https://doi.org/10.1080/00218464.2023.2300453
10. Friedman, H.L.: Kinetics of thermal degradation of char-forming plastics from thermogravimetry. Application to a phenolic plastic. J. Polymer Sci. Part C: Polymer Symposia **6**(1), 183–95 (1964). https://doi.org/10.1002/polc.5070060121
11. Mashouf Roudsari, G., Mohanty, A.K., Misra, M.: Study of the Curing Kinetics of Epoxy Resins with Biobased Hardener and Epoxidized Soybean Oil. ACS Sustain. Chem. Eng. **2**(9), 2111–6 (2014). https://doi.org/10.1021/sc500176z
12. Zhang, J., Wang, X., Zhang, S., Gao, Q., Li, J.: Effects of Melamine Addition Stage on the Performance and Curing Behavior of Melamine-Urea-Formaldehyde (MUF) Resin. BioResources 8 (2013). https://doi.org/10.15376/biores.8.4.5500-5514
13. Lechner, M.D.: Einführung in die Kinetik. Springer (2018)
14. Michaelides, A., et al.: Identification of General Linear Relationships between Activation Energies and Enthalpy Changes for Dissociation Reactions at Surfaces. J. American Chem. Soc. **125**(13), 3704–5 (2003). https://doi.org/10.1021/ja027366r
15. Pizzi, A., Mittal, K.: Urea-formaldehyde adhesives. Handbook of adhesive technology. V2 (2003)
16. Tomita, B., Hirose, Y.: Urea-formaldehyde resins: NMR study on base-catalyzed reaction of formaldehyde with urea in deuterium oxide. J. Polymer Sci. Polymer Chem. Edit. **14**(2), 387–401 (1976). https://doi.org/10.1002/pol.1976.170140211
17. Jong, J., Jonge, J.: Kinetics of the formation of methylene linkages in solutions of urea and formaldehyde. Recl. Trav. Chim. Pays-Bas **72**, 139–156 (1952). https://doi.org/10.1002/recl.19530720211
18. Jong, J., Jonge, J.: Kinetics of the reaction between mono-methylolurea and methylene diurea. Recueil des Travaux Chimiques des Pays-Bas. 72 (1952). https://doi.org/10.1002/recl.19530720306
19. Jong, J., Jonge, J.: The reaction of methylene diurea with formaldehyde. Recueil des Travaux Chimiques des Pays-Bas. 72 (1952). https://doi.org/10.1002/recl.19530720307
20. Jong, J., Jonge, J.: The reaction of urea with formaldehyde. Recl. Trav. Chim. Pays-Bas **71**, 643–660 (1951). https://doi.org/10.1002/recl.19520710704
21. Nair, B.R., Francis, D.J.: Kinetics and mechanism of urea-formaldehyde reaction. Polymer **24**(5), 626–30 (1983). https://doi.org/10.1016/0032-3861(83)90118-0
22. Erä, V.A., Lindberg, J.J., Mattila, A., Vauhkonen, L., Linnahalme, T.: Phenol-formaldehyde impregnant characterization by infrared spectroscopy and differential scanning calorimetry. Die Angewandte Makromolekulare Chemie **50**(1), 43–52 (1976). https://doi.org/10.1002/apmc.1976.050500105
23. Holopainen, T., Alvila, L., Rainio, J., Pakkanen, T.T.: IR spectroscopy as a quantitative and predictive analysis method of phenol–formaldehyde resol resins. J. Appl. Polymer Sci. **69**(11), 2175–85 (1998). https://doi.org/10.1002/(SICI)1097-4628(19980912)69:11<2175::AID-APP9>3.0.CO;2-D
24. Poljansek, I., Krajnc, M.: Characterization of phenol-formaldehyde prepolymer resins by in line FT-IR spectroscopy. Acta Chim. Slov. **52**(3), 238 (2005)

25. Siimer, K., Kaljuvee, T., Christjanson, P.: Thermal behaviour of urea-formaldehyde resins during curing. J. Thermal Analy. Calorim. **72**(2), 607–17 (2003). https://doi.org/10.1023/A:1024590019244
26. Szesztay, M., László-Hedvig, Z., Kovacsovics, E., Tüdős, F.: DSC application for characterization of Urea/formaldehyde condensates. Holz als Roh- und Werkstoff **51**(5), 297–300 (1993). https://doi.org/10.1007/BF02663798
27. Siimer, K., Kaljuvee, T., Christjanson, P., Lasn, I.: Curing of urea-formaldehyde resins on awood substrate. J. Thermal Analy. Calorim. **84**(1), 71–7 (2006). https://doi.org/10.1007/s10973-005-7185-7
28. Siimer, K., Kaljuvee, T., Christjanson, P., Pehk, T.: Changes in curing behaviour of aminoresins during storage. J. Thermal Analy. Calorim. **80**(1), 123–30 (2005). https://doi.org/10.1007/s10973-005-0623-8

Thermal Characterization of Adhesives and Studies of CTR Between Metal and Adhesives

Serges Kikwani[1(✉)], Nathalie Ehret[2], Jean-Laurent Gardarein[2], and Fréderic Lebon[1]

[1] Aix Marseille Université, LMA, CNRS, CEDEX 13, 13453 Marseille, France
{kikwani,lebon}@lma.cnrs-mrs.fr

[2] Aix Marseille Université, IUSTI, CNRS, CEDEX 13, 13453 Marseille, France
{nathalie.ehret,jean-laurent.gardarein}@univ-amu.fr

Abstract. The aim of this study is to evaluate the contact thermal resistance (CTR) between the metal and structural adhesives in the crystalline state, along the axis of revolution of the 3D model of the specimen, during a thermal transfer in the time regime (at t = 600 s). As the data sheets of the adhesives studied do not provide all the thermophysical properties, the first step is to characterize these adhesives experimentally. In other words, to determine thermal conductivity, density, specific heat and thermal diffusivity required to solve the heat transfer equation in the 3D model of the specimen. The second step is to determine the temperature field along the axis of revolution of the 3D specimen model, using the 3D finite element method (FEM) with COMSOL Multiphysics software. The third step involves evaluating the CTR between metal and structural adhesive at two interfaces (metal-adhesive and adhesive-metal), based on temperatures (T_1, T_2, T_3, T_α and T_β) obtained numerically. The results of this work show that at the M interface (metal-adhesive contact), the CTR is low, while it is high at the N interface (adhesive-metal contact).

Keywords: Structural adhesive · Density · Thermal conductivity · Thermal diffusivity · Specific heat · Porosity · CTR

1 Introduction

Over recent decades, structural bonding technologies have taken a significant leap in the aerospace field (A. Maurel-Pantel et al., 2022), thanks to advances in the development of high-performance polymeric materials and the need for lighter structures (M. Lamberti et al., 2016). Structural bonding is a suitable assembly method for the composite materials usually used for these structures. Recent advances in this field have led to their use in the automotive (L.D.R. Grant et al., 2009), naval or marine (K. Idrissa et al., 2024, A. Maurel-Pantel et al., 2022, R. Niu et al, 2023) and civil engineering sectors (A. Billon, 2016). This work aims to theoretically evaluate the temperature profile along the axis of revolution of a 3D model of a specimen consisting of two metal substrates bonded with a structural adhesive. It also aims to assess the contact thermal resistance (CTR) at two

L. F. M. da Silva and R. D. Adams (Eds.): AB 2025, PEM, pp. 33–50, 2026.
https://doi.org/10.1007/978-3-032-19022-2_3

interfaces of this specimen during heat transfer over a period of 600 s, using the 3D finite element method with COMSOL Multiphysics software (COMSOL). Since the equations describing heat transfer between two solids in contact have been well identified in many works, these equations are only applicable to perfect contacts (R. Chadoul, 2018). But in reality, perfect contact between two solids does not exist. Experimentally, a sudden drop in temperature can be observed at the interface between the two solids. This phenomenon has led to the idea of CTR. However, the fact that contact surfaces are not perfectly flat at the microscopic level leads us to consider three types of thermal diffusion: conduction, convection and radiation. The contact zone is a structural discontinuity in the materials involved. In fact, according to some authors (R. Chadoul et al., 2017, R. Chadouli et al., 2019), contact takes place only on a relatively small number of points, with the effective surface area of real contact compared to the apparent surface area representing only 1 to 3% and no more than 6 to 8% for well-polished surfaces. This phenomenon leads to CTR at the interface, which is influenced by asperity size and the interstitial fluid, which is generally a poor conductor (R. Chadoul, 2018, G. Kaza, 2010).

Several studies have been conducted on the CTR between two metal, and we will mention just a few of them. Chadouli et al., 2017 presented and then analytically validated a 2D steady-state numerical model to study the impact of the thermal quality of the joint on the CTR between two solids. Chadouli et al., 2019 carried out a numerical simulation using the ABAQUS finite element code to study the influence of thermal conductivity, the interstitial fluid convection coefficient, and surface roughness on the CTR, with the aim of improving heat transfer between materials in contact. R. Dou et al., 2016 developed an experimental setup to study the effects of interface temperature, sample surface roughness, and contact pressure on contact thermal conductance (CTC). S. C. Some et al., 2013 have conducted a comparative and validation study of prediction models for CTR in the solid–liquid case.

As this study aims to evaluate the CTR between metal and adhesive in a crystalline state, six structural adhesives were selected based on availability: four epoxy adhesives, one polyurethane adhesive, and one methacrylate adhesive. Regarding metals, aluminium, copper and steel were chosen. To evaluate the CTR, it is necessary to understand the geometry, the thermophysical properties, and the exchange mechanisms at the interface. While a relatively small change in structure can be correlated with a significant change in the value of specific heat, the other two parameters—thermal conductivity and thermal diffusivity—are transport parameters linked to the material's ability to conduct heat.

The data sheets do not provide information on all the thermophysical adhesive properties as shown in Table 2, it was decided to characterise them. However, a brief bibliographical study was first carried out. B. Suleiman, 2018, who studied thermophysical properties directly related to thermal conduction, such as the thermal conductivity and thermal diffusivity of composite polymer materials (polystyrene and polymethyl methacrylate (PMMA). H. Lobo, 2003, who presented thermal conductivity data as a function of temperature for five commercial polymers (three amorphous and two semi-crystalline).

The results of this work indicate that the values obtained during the thermophysical characterization tests of the adhesives are satisfactory and consistent with some of the values indicated in the technical data sheets. With regard to the CTR study, the copper-adhesive contact has a low CTR at the M interface, while the steel-adhesive contact has a low CTR at the N interface, with the exception of the Sicomin adhesive.

2 Materials

2.1 Studied Metals

Three metals were chosen for this work: aluminium 3003-H18, structural steel, and copper. Their thermophysical properties are shown in the Table 1.

Table 1. Thermo-physical properties of the various metals used [COMSOL]

Specification	Aluminium	Steel	Copper
Density $\rho\left(kg/m^3\right)$	2,730	7,850	8,960
Specific heat $C_p(J/kg.K)$	893	470	385
Thermal conductivity $\lambda(W/m.K)$	155	50.2	400

2.2 Studied Adhesives

The following adhesives were chosen for this work: Loctite UK 1351B25, Saf 30 MIB, Sicomin SR 5030, Araldite AV138M-1/HV998, 3M Scotch Weld 2216 B/A and Sikadur 30. Table 2, summarises the thermophysical properties of the adhesives, as detailed in the data sheets.

Table 2. Thermo-physical properties of the various adhesives from the data sheets

Specification	Loctite	Saf 30	Sicomin	Araldite	3M2216	Sikadur 30
Classification of various Adhesives studied						
Type	Polyurethane	Methacrylate	Epoxy	Epoxy	Epoxy	Epoxy
Manufacturing Company	Henkel	Bostik	Sicomin	Huntsman	3M Company	Sika
Thermo-physical properties of the various adhesives studied						
Density $\rho\left(kg/m^3\right)$	1,410	1,200	1,250	-	-	-
Specific heat $C_p(J/kg.K)$	-	-	-	-	-	-

(*continued*)

Table 2. (*continued*)

Specification	Loctite	Saf 30	Sicomin	Araldite	3M2216	Sikadur 30
Thermal diffusivity $a\left(10^{-6}m^2s^{-1}\right)$	-	-	-	-	-	-
Thermal conductivity $\lambda(W/m.K)$	-	-	-	-	0.39460758	-

(-) mean, Missing value indicate in the data sheet.

2.3 Manufacturing of Molds and Preparation of Adhesives

The Sample molds were designed using SolidWorks (Systèmes, 2025) [5], then manufactured by 3D printing with a Raise 3D N2 Plus Dual printer (Raise 3D, China), using 1.75 mm diameter PLA + (Polylactic acid) plastic filaments. Table 3 lists the dosages, curing times and mixing method used to prepare the adhesives. Adhesives were prepared in a conditioned room, at a controlled temperature of 23 °C and relative humidity of (50 ± 5) %.

Table 3. Composition, mixing methods, curing conditions, and resin-to-hardener ratios for the preparation of the studied adhesives from the data sheets

Specification	Loctite	Saf 30	Sicomin	Araldite	3M2216	Sikadur 30
Resin	UK 1351 B25	Comp A	Isobon SR 5030	AV 138M-1	2216 - Comp B	Sika - Comp A
Hardener	UK 5452	Comp B	Isobon SD 5032	HV998	2216 - Comp A	Sika - Comp B
Mixing Method	Gun	Gun	Manuel	Manuel	Manuel	Manuel
Curing Time	24h at 23 °C	24h at 23 °C	-	24h at 23 °C	7 days at 24 °C	48h at 23 °C
Resin/Hardener Ratio	10/4 g	2/1 ml	10/4.5 g	10/4 g	5/7 g	4.5/1.5 g

(-) mean, Missing value indicate in the data sheet.

2.4 3D Model of Specimen

For the numerical study, a bonding model of a specimen consisting of two metal substrates, bonded with an adhesive joint, is proposed, as shown in Fig. 1. This specimen has two metal-adhesive interfaces M and N, three 9 mm deep holes spaced 6 mm apart (one on the upper substrate, one in the adhesive joint and one on the lower substrate). These holes are located 3 mm from the two interfaces (e = δ = 3 mm).

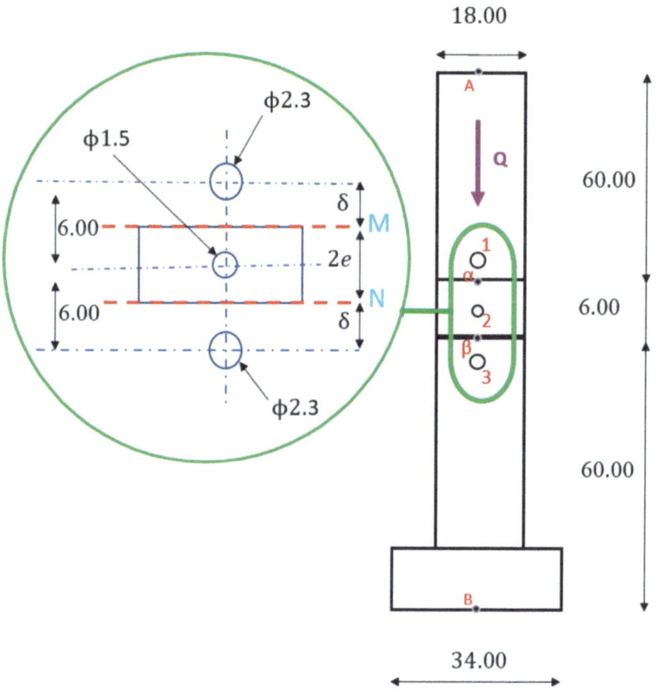

Fig. 1. Dimensions of the Specimen (mm)

where: 1 is the hole in the upper substrate, 2 is the hole in the lower substrate, 3 is the hole in the adhesive joint, $2 * e$ is the thickness of adhesive joint, M is the Interface between the upper substrate and the adhesive, N is the Interface between the adhesive and the lower substrate, α is the intersection point between the specimen's revolution axis and the M interface, β is the intersection point between the specimen's revolution axis and the N interface, δ is the distances between the interfaces and holes (1 and 2) and Q is the heat flux.

3 Experiment Tests: Thermal Characterization of Adhesives

The aim of this section is to determine some of the thermophysical properties of adhesives needed to solve the heat equation. These are thermal conductivity k, thermal diffusivity a, specific heat Cp and density ρ. The Table 4 shows the various dimensions of the adhesive samples. For the various types of test, the mean values of the measurements on the different samples were retained and presented. As the measuring equipment was accurate, three measurements on each sample were sufficient to determine the relative deviations E_r of these measurements. These relative deviations were calculated using

$$E_r(\%) = \left[\frac{2(x_{max} - x_{min})}{x_{max}}\right] * 100 \tag{1}$$

Table 4. Various dimensions of samples

Specification	Density Test	Conductivity Test	Diffusivity Test	Specific Heat Test
Shape of the sample	Square	Parallelopiped	Square	Cylinder
Dimension of the sample (mm)	7 × 7	70 × 40	20 × 20	4
Thickness or length (mm)	8.4	10	2	100
Number of samples per adhesive	3	2	3	3

3.1 Density Test

The density measurements were carried out at atmospheric temperature, using a vernier caliper (Apparent density ρ_{app}) and using a helium pycnometer and a densitometer (Real density ρ_s).

3.1.1 Apparent Density Measurement

The apparent density ρapp of various samples were evaluated using balance for (Scientech Balance ZSA 120, Max: 120g - Min: 0.01g, precision d = 0.0001 g) the mass m_{app}, , and calliper (Mitutoyo micrometer, Max: 25mm - Min: 0, precision: 0.001mm) for the apparent volume V_{app} by

$$\rho_{app} = \frac{m_{app}}{V_{app}} \tag{2}$$

3.1.2 Density Test Using Accupyc II 1340 Helium Pycnometer (Micromeritics Instrument Corporation, USA)

The density of the adhesive samples using Accupyc II 1340 Helium pycnometer in accordance with ASTM D4892-24 Standard, is the non-destructive technique uses helium's ability to penetrate nanometric pores to measure volume by gas displacement. Based on the law of perfect gases, it compares the pressure in a chamber containing the sample with that in an empty reference chamber. After filling a known volume with helium Vcel, it's released into an expansion chamber of volume V_{exp}. The resulting change in pressure is then used to estimate volume of sample V_s by

$$V_s = \frac{V_{cel} - V_{exp}}{\left(\frac{p_1}{p_2}\right) - 1} \tag{3}$$

The pressure p_1 is controlled when the sample chamber (whose volume V_{cel} is known) is filled. The chamber is then opened to a second expansion chamber with a pressure of p_2 and a volume of V_{exp}. The instrument measures the volume of the sample, excluding interstitial voids and open porosity accessible to gas in individual particles. Closed porosity, is included in the volume. The density was determined using

$$\rho_{app} = \frac{m_{app}}{V_s} \tag{4}$$

3.1.3 Density Test Using OHAUS Densitometer (Nänikon, Switzerland)

The density of the adhesive samples was evaluated OHAUS densitometer in accordance with ISO 1183 Standard using

$$\rho_s = \frac{m_s}{m_s - m_d}(\rho_d - \rho_a) + \rho_a \tag{5}$$

where ρ_d is the density of distilled water and ρ_a is the density of air, m_s represents the mass of the sample measured in the open air, and m_d its mass after immersion in a distilled water bath. m_s and m_d were calculated from the average of five measurements to ensure accurate results. As the densitometer is not as accurate as the pycnometer, 10 measurements were taken on each sample. Relative dispersion CV is calculated using

$$CV = \frac{Standard\ deviation}{Meam} \times 100 \tag{6}$$

3.1.4 Deduced Porosity

The deduced porosity of adhesives ϕ is estimated using

$$\phi = \frac{\rho_s - \rho_{app}}{\rho_s} \tag{7}$$

3.2 Specific Heat Test (TIAN-CALVET, France)

The specific heat of the various adhesive samples was determined using a TIAN - CALVET drop calorimeter, in accordance with ISO 11357–4 and ASTM E1269 Standards. The specific heat was measured by dropping sample for an average temperature around 50 °C (initially at room temperature 29 °C and 74 °C).

3.3 Thermal Diffusivity Test: Laser Flash Analysis (LFA 1000) (Linseis, Germany)

The thermal diffusivity of the adhesives was measured using the LFA 1000 equipment, in accordance with the standards ASTM E-1461, DIN 30905 and DIN EN 821.

3.4 Thermal Conductivity

3.4.1 Deduced Thermal Conductivity

The deduced thermal conductivity of adhesives were calculated using

$$k(T) = a(T)\rho(T)C_p(T) \tag{8}$$

3.4.2 Transient Hot Bridge (THB) Method (Linseis, Germany)

The THB100 equipment complies with the following ASTM: D5930–01, ASTM D7896–19 and ASTM C1113 standards, as well as ISO 22007–2, it measures thermal conductivity using a transient method.

3.4.3 Hot Disk Method (Thermoconcept, Germany)

Hot Disk equipment, which complies with ISO 22007–2 standard, measures thermal conductivity using olso a transient technique, the measured samples are the same as those used with the THB 100 method.

4 The 3D Model of Numerical FEM Analysis of the Specimen: CTR Between Metal – Adhesive

The numerical studies carried out in this work are based on a temporel 3D FEM at 600 s, using COMSOL Multiphysics. The top surface of the specimen is heated to 200 °C (473.15K), while the bottom surface of the specimen and side surfaces are assumed to be at 20 °C (293.15K). The 3D mesh is show in Fig. 2a, the statistical specifications and size of mesh are shown in Tables 5 and 6.

Table 5. Mesh statistics with COMSOL Multiphysics software

Mesh Statistics							
Mesh Points	Tetrahedra elements	Triangles	Edge elements	Point elements	Number of elements	Minimum element quality	Average element quality
6,099	28,867	5,129	489	26	28,867	0.008329	0.6512

Table 6. Mesh size with COMSOL Multiphysics software

Mesh Size					
Maximum element size	Manimum element size	Curvature Factor	Resolution of narrow regions	Maximum element growth rate	Predefined Mesh size
0.00441	1.89×10^{-4}	0.3	0.85	1.35	Extra fine

The heat transfer model in solids uses the version of the heat equation in Eq. 9 as the mathematical model.

$$\rho C_p \frac{\partial T}{\partial t} - \nabla.(\lambda \nabla T) = Q \tag{9}$$

where: T represents the temperature field within the 3D model of the test specimen and t the time.

4.1 Temperature Evolution Along the Axis of Revolution

To visualize the temperature profile within the specimen, the 3D Cutting line post-processing tool in COMSOL Multiphysics (COMSOL Inc, 2022) was applied as show along the axis of revolution, as illustrated in Fig. 2b.

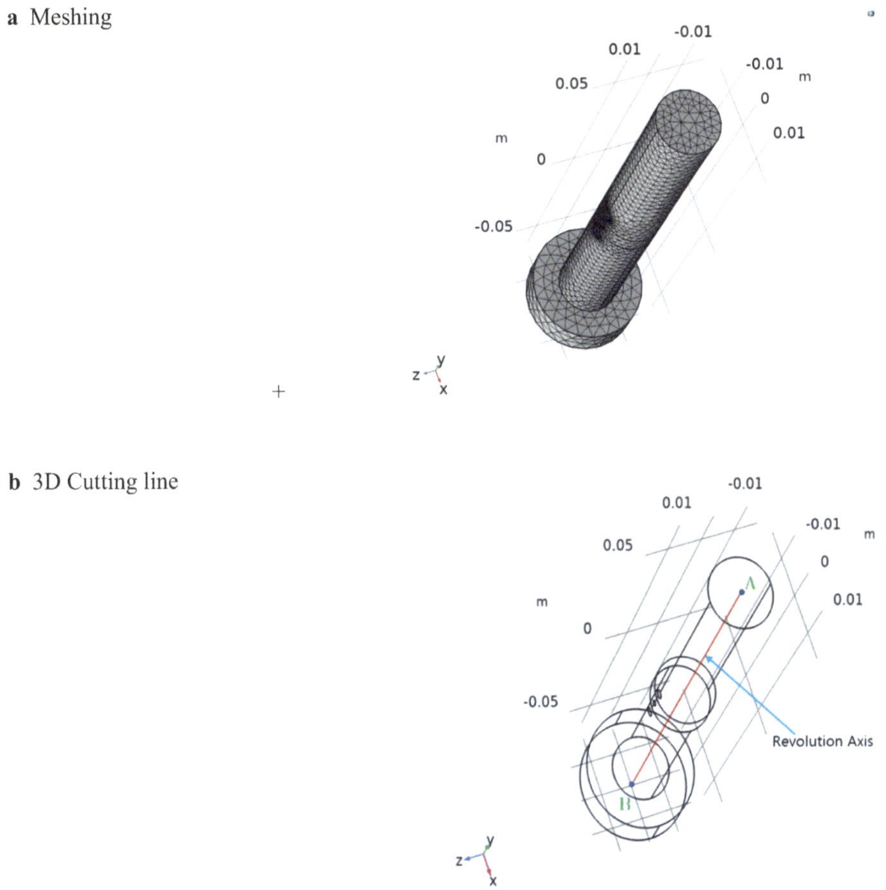

Fig. 2. Meshing and 3D Cutting line with COMSOL

4.2 CTR Between Metal - Adhesive at Interface M and CTR Between Adhesive - Metal at Interface N

In this work, the CTR is evaluated on the basis of the temperatures of the three holes in the specimen, at two interfaces of a specimen composed of two metals joined by an adhesive joint. Given that a heat input was imposed at the top of specimen, in contrast to the work of authors (R. Chadouli et al., 2019) where two heat inputs were imposed, at the top of the upper specimen and at the bottom of the lower specimen as well as in contrast to the work of authors (R. Dou et al., 2016), where a heat input was imposed at the bottom of the lower specimen and a cooling was imoposed at the top of the upper specimen.

For smooth surfaces, the contact ratio between two solids does not exceed 10% of the nominal contact area (R. Chadoul, 2018). Under pressure, The experimental contact ratio between the metal and the amorphous adhesive could range from 5% to 30%, which is significantly higher than the contact ratio for metal-to-metal contact. This theoretical work focuses on the contact between the metal and the crystalline adhesive, the contact ratio between them is assumed to be 20%.

4.2.1 CTR at Interface M

The CTR at interface M is given using

$$CTR_{\alpha} = \frac{T_1 - T_2}{q_M} \tag{10}$$

The heat flux across to the M interface is given by

$$q_M = \frac{\lambda S_M}{e}(T_1 - T_2) \tag{11}$$

With Eq. 10 and Eq. 11, the expression becomes:

$$CTR_{\alpha} = \frac{2e(T_1 - T_2)}{0.1\lambda\pi D^2(T_1 - T_{\alpha})} \tag{12}$$

where: λ is the thermal conductivity of the metal and D is the diameter of the specimen at the adhesive joint.

4.2.2 CTR at Interface N

The CTR at interface N is given using

$$CTR_{\beta} = \frac{T_2 - T_3}{q_N} \tag{13}$$

The heat flux across the N interface is given by

$$q_N = \frac{\lambda S_N}{\delta}(T_2 - T_{\beta}) \tag{14}$$

With Eq. 13 and Eq. 14, the expression becomes:

$$CTR_{\beta} = \frac{2\delta(T_2 - T_3)}{0.1k\pi D^2(T_2 - T_{\beta})} \tag{15}$$

where: k is the thermal conductivity of the adhesive.

5 Results

5.1 Experiment Results

5.1.1 Density and Porosity of the Various Adhesive Samples Studied

For various adhesive, the results of the density are shown in Fig. 3 and the results of porosity are shown in the Table 7.

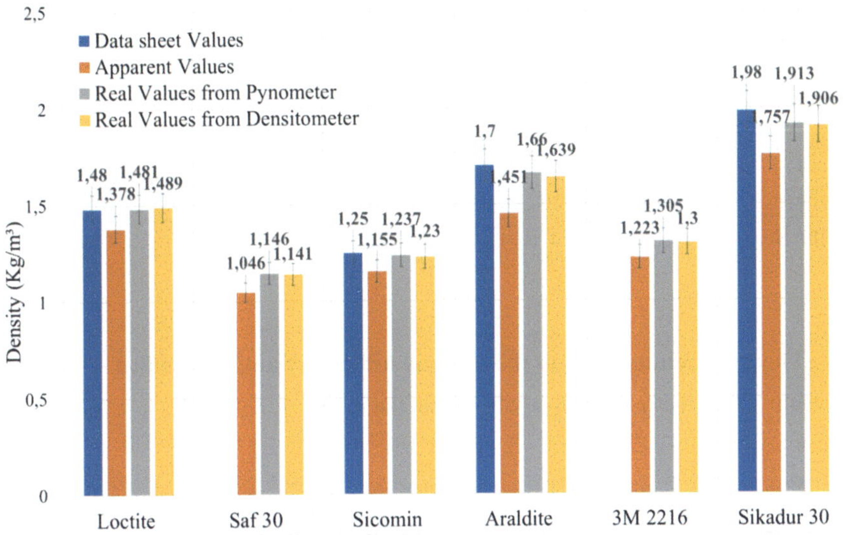

Fig. 3. Measured densities of the six adhesives investigated, along with reference values

Table 7. Apparent porosity of various Ahesive studied

Specification	Loctite	Saf 30	Sicomin	Araldite	3M2216	Sikadur 30
Porosity from pycnometer ϕ_{pyc} (%)	6.9	8.7	6.6	7.2	6.3	7.9
Porosity from densitometer ϕ_{den} (%)	7.5	8.3	6.1	6	6	7.8

5.1.2 Thermal Diffusivity of the Various Adhesive Samples Studied

The results of thermal diffusivity of adhesives are shown in Table 8.

Table 8. Measured Thermal diffusivity of the various adhesive samples investigated

Specification	Loctite	Saf 30	Sicomin	Araldite	3M2216	Sikadur 30
Thermal diffusivity $a\left(10^{-6}m^2s^{-1}\right)$	0.21	0.11	0.15	0.18	0.17	0.91

5.1.3 Specific Heat of the Various Adhesive Samples Studied

The results of specific Heat are shown in Table 9.

Table 9. Measured Specic Heat of the various adhesive samples investigated

Specification	Loctite	Saf 30	Sicomin	Araldite	3M2216	Sikadur 30
Specific Heat $C_p(J/kg.K)$	1,234	1,767	1,308	798	1,734	939

5.1.4 Thermal Conductivity of the Various Adhesive Samples Studied

The results of the thermal conductivity are shown in Fig. 4.

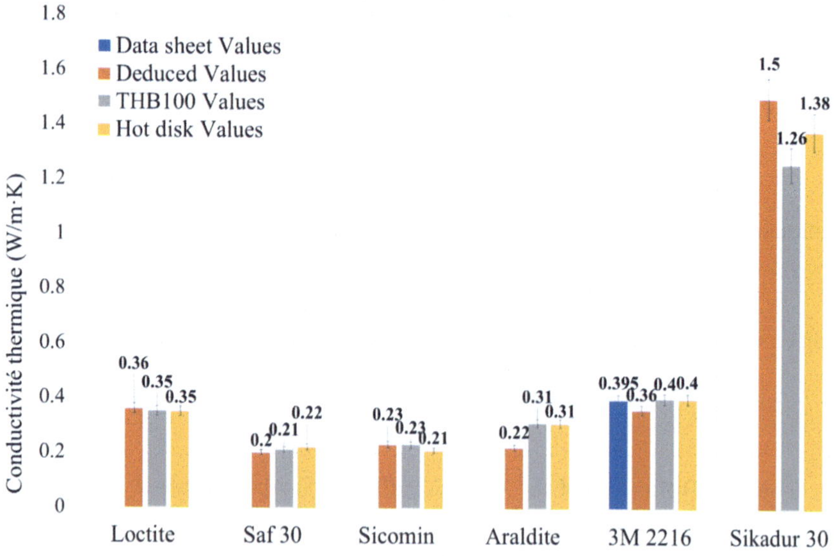

Fig. 4. Measured Thermal conductivity of the six adhesives investigated, along with one reference values reported in manufacturer datasheets

5.2 The 3D Model of Numerical FEM Analysis of the Specimen: CTR Between Metals and Adhesive

5.2.1 The Temperature Evolution in the Specimen

The results of 3D Field of Temperature on the specimen is shown in Fig. 5 and the 3D temperature fields plotted along the specimen's axis of revolution are shown in Fig. 6.

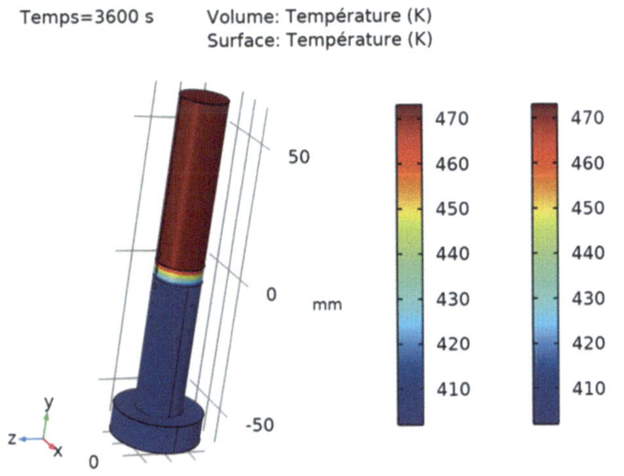

Fig. 5. 3D Field of Temperature on the Alu-loctite adhesive specimen using the 3D finite element method (FEM) with COMSOL Multiphysics software (COMSOL Inc, 2022)

a Copper - Adhesives

b Aluminium - Adhesives

c Steel - Adhesives

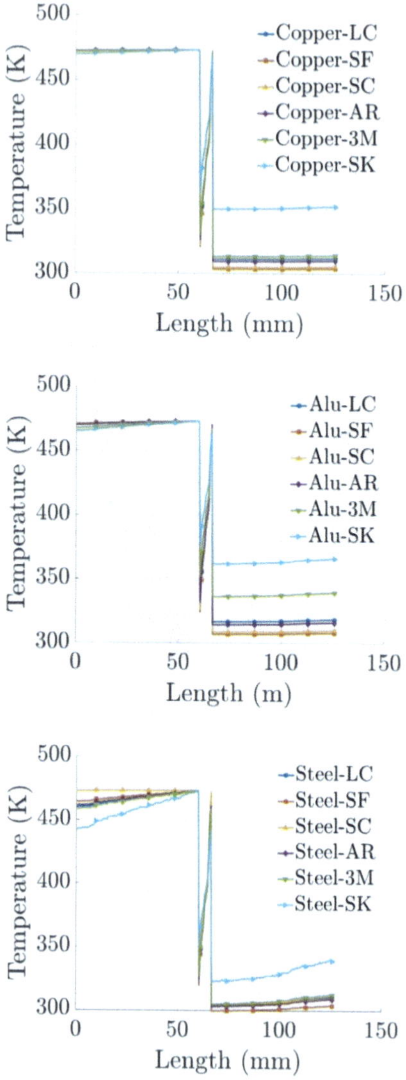

Fig. 6. Temperature jump profile within the specimen by 3D FEM using COMSOL

5.2.2 CTR Between Metal - Adhesive and CTR Between Adhesive – Metal

The results of the CTR between metal and adhesive (CTR_{α}) and between adhesive and metal (CTR_{β}) are shown in Figs. 7 and 8.

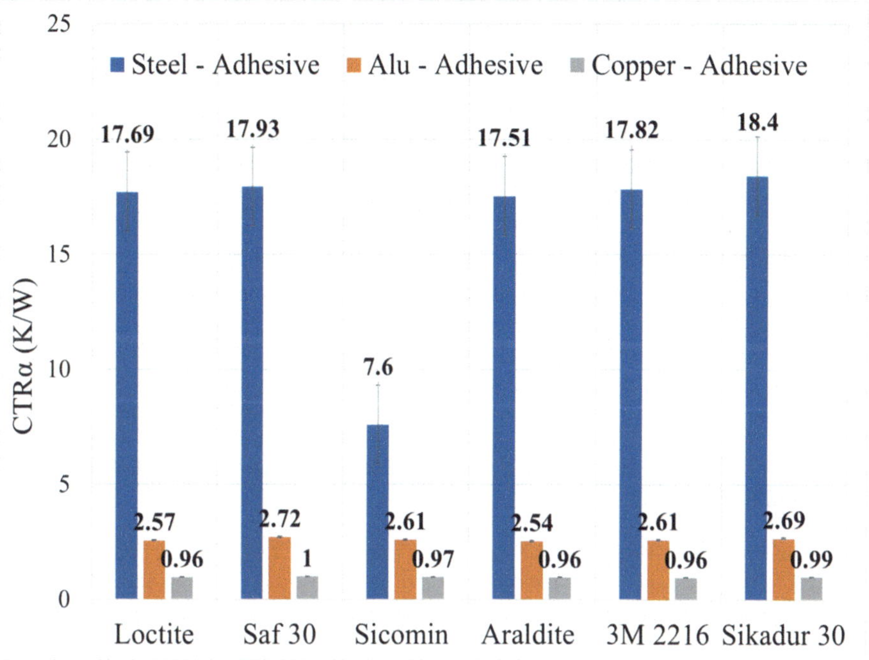

Fig. 7. CTR between metal and adhesive at M interface

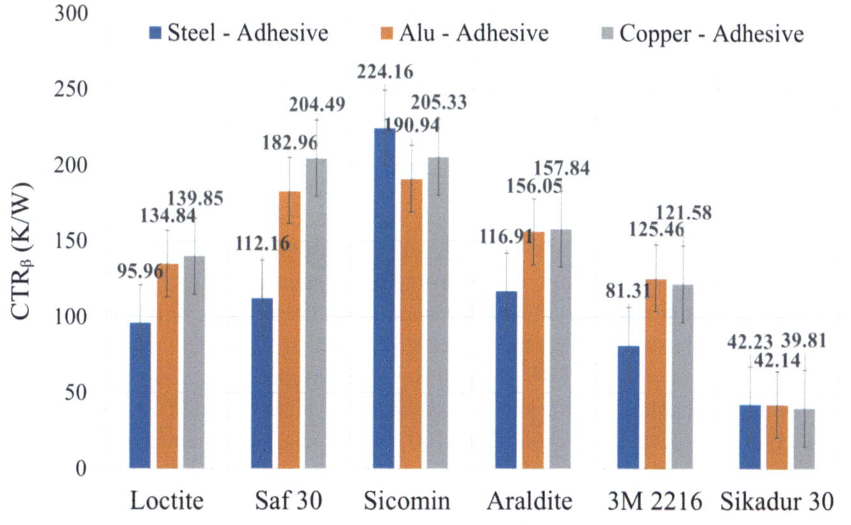

Fig. 8. CTR between adhesive and metal at N interface

6 Discussion

6.1 Thermal Characterisation of Various Adhesive Studied

Based on the results in Fig. 3, the dispersion of apparent density measurements is 4%, the dispersion of real density measurements is 1.5% with the densitometer and 4% with the pycnometer. Sikadur 30 is the densest. Based on the specific heat results shown in Table 9, the dispersion is around 8%. SAF 30 and 3M 2216 adhesives have a higher specific heat capacity than other adhesives. Based on the thermal diffusivity results, the dispersion is around 7%, Sikadur 30 adhesive is the most diffusive. As can be seen from the results in Fig. 4, the thermal conductivity results, the dispersion is around 10% for deduced values. For measurements carried out on THB100, the dispersion is 4%, and 5% for measurements carried out on Hotdisk. Sikadur 30 adhesive is the most conductive.

6.2 The 3D Model of Numerical FEM Analysis of the Specimen: CTR Between Metal and Adhesive

As can be seen from Figs. 7 and 8, at the M interface, CTR_M is low, while at the N interface, CTR_N is very high. This can be explained by the fact that the ratio of the thermal conductivity of the metal to that of the adhesive ($r_M = \lambda/k$) is significantly greater than 1 at the M interface, while the ratio of the thermal conductivity of the adhesive to that of the metal ($r_N = k/\lambda$) is less than 1 at the N interface. However, the copper-adhesive contact at the M interface offers a low CTR, while the adhesive-steel contact at the N interface also offers a low CTR, with the exception of Sicomin adhesive.

7 Conclusion and Perspectives

This work enabled the CTR at the interface between metal and somes adhesives to be evaluated. Experimental tests were successfully carried out to determine the thermo-physical properties of the adhesives required for the heat equation within the specimen model, namely density, thermal conductivity, mass heat capacity and thermal diffusivity. The experimental results obtained on the adhesive samples from the pycnometer and densitometer are in good agreement with some of the density values given in the data sheets, validating the reliability of the measurements. Conversely, only the data sheet for the 3M2216 adhesive provides a conductivity value that aligns with the experimental measurements obtained using THB 100 and Hot Disk equipment, as well as the deduced value. For the other adhesives, the various conductivity values obtained from the samples (i.e. the experimental values from the THB 100 and Hot Disk equipment and the deduced values) are consistent, except for the Sikadur 30 adhesive, for which the deduced value shows a slight discrepancy.

The CTR was evaluated using temperature profiles plotted along the axis of revolution of the specimen model through the 3D FEM with COMSOL Multiphysics software. The M interface exhibited a lower CTR than the N interface for all metal-adhesive couplings. Copper-adhesive couplings showed low CTR at the M interface, while steel-adhesive couplings showed low CTR at the N interface, except for the steel-Sicomin coupling.

Since the CTR values were evaluated on the basis of the temperature field calculated numerically along the axis of revolution of the specimen containing holes close to the interfaces and whose depths coincide with this axis, these values therefore depend on the properties of the materials in contact, the temperature field and the depth of the holes. To complete this work, it is possible to experimentally determine the actual contact ratio between the metal and the adhesive under the same conditions, and to test the CTR between the two materials.

Acknowledgments. Serges Kikwani expresses his sincere gratitude to the administrators of the French Government Scholarship Program (BGF) for their generous financial support.

References

Billon, A.: Méthode d'évaluation non-destructive de la qualité du collage des composites de renforcement pour le génie civil. PhD thesis. Université Paris-Est (2016)

Chadoul, R.: Analyse des Phénomènes mécaniques aux interfaces sous charge et leurs effets sur la RTC, PhD thesis. Université Djillali Liabes (2018)

Chadouli, R., Lebon, F., Rosu, I., Makhlouf, M.: Numerical study of the gasket thermal conductivity effect on the thermal contact resistance between two solids in contact. Frontiers in Heat and Mass Transfer **8**(30) (2017)

Chadouli, R., Lebon, F., Rosu, I., Makhlouf, M.: Numerical study of the surface roughness, thermal conductivity of the contact materials and interstitial fluid convection coefficient effect on the thermal contact conductance. In Annales de Chimie-Science des Matériaux **43**, 265–271 (2019)

Systèmes, D.: SolidWorks Simulation Software (2025). https://www.solidworks.com/. Version used : SolidWorks 2025

Dou, R., Ge, T., Liu, X., Wen, Z.: Effects of contact pressure, interface temperature, and surface roughness on thermal contact conductance between stainless steel surfaces under atmosphere condition. Int. J. Heat Mass Transf. **94**, 156–163 (2016)

Grant, L.D.R., Adams, R.D., da Silva, L.F.M.: Effect of the temperature on the strength of adhesively bonded single lap and t joints for the automotive industry. Int. J. Adhesion and Adhesives **29**(5), 535–542 (2009)

Idrissa, K., Maurel-Pantel, A., Lebon, F., Guermazi, N.: On the durability performance of two adhesives to be used in bonded secondary structures for offshore wind installations. Materials **17**(10), 2392 (2024)

COMSOL Inc. Comsol multiphysics, version 6.1.0.357 (2022)

Kaza, G.: Contribution à l'étude de la résistance thermique de contact et à sa modélisation à travers l'écrasement de l'interface tôle/outil dans la mise en forme à chaud de tôles d'acier, PhD thesis. Université de Toulouse, Université Toulouse III-Paul Sabatier (2010)

Lamberti, M., Maurel-Pantel, A., Ascione, F., Lebon, F.: Influence of web/flange reinforcement on the gfrp bonded beams mechanical response : A comparison with experimental results and a numerical prediction. Compos. Struct. **147**, 247–259 (2016)

Lobo, H.: Thermal conductivity and diffusivity of polymers. In: Handbook of plastics analysis, pp. 131–154. CRC Press (2003)

Maurel-Pantel, A., Voisin, M., Mazerolle, F., Lebon, F.: Comparison of three different adhesive joints using static and dynamic impact tests: Development of a new drop weight impact test rig incorporating a modified arcan fixture. Int. J. Adhes. Adhes. **114**, 103104 (2022)

Niu, R., Yang, Y., Liu, Z., Ding, Z., Peng, H., Fan, Y.: Durability of two epoxy adhesive bfrp joints dipped in seawater under high temperature environment. Polymers **15**(15), 3232 (2023)

Some, S.C., Delaunay, D., Gaudefroy, V.: Étude comparative et validation de modèles prédictifs de résistance thermique de contact dans le cas solide-liquide avec prise en compte de la tension superficielle. In: CFM 2013-21ème Congrès Français de Mécanique. AFM, Maison de la Mécanique, 39/41 rue Louis Blanc-92400 Courbevoie (2013)
Suleiman, B.M.: Thermal conductivity and diffusivity of polymers. Int. J. Phys. Study Res. **1**, 54–58 (2018)

Optimizing Green Strength in Polyurethane Adhesives via TPU Incorporation and Controlled Cooling of the Adhesive Joint

María Pilar Carbonell-Blasco(✉), Alejandra Moyano-Vallejo, Carlota Hernández-Fernández, María Dolores Romero-Sánchez, and Elena Orgilés-Calpena

INESCOP Footwear Technology Centre, C/ Alemania 102, 03600 Elda, Alicante, Spain
{pcarbonell,amoyano,chernandez,mdromero,eorgiles}@inescop.es

Abstract. Reactive polyurethane hot-melt adhesives (HMPURs) are widely used in industrial applications due to their combination of thermoplastic and thermosetting properties, providing strong adhesion and long-term stability. However, green strength remains a critical factor in fast-paced manufacturing processes. This study investigates the effect of incorporating thermoplastic polyurethane (TPU) and the impact of rapid cooling after bonding on the adhesion performance, mechanical properties and thermal stability of polyester-polyether-based HMPURs. Three formulations were synthesized: a reference HMPUR and two TPU-modified adhesives (HMPUR-10ECO and HMPUR-15ECO) containing 10 and 15 wt% TPU, respectively. The adhesives were characterized using FTIR, TGA, DSC, rheology and DMTA to evaluate their structural, thermal and viscoelastic properties. T-peel strength tests were conducted under standard conditions and after rapid cooling. The results show that TPU incorporation significantly enhances initial adhesion, particularly when combined with rapid cooling. The cooling process accelerates the solidification of the adhesive, improving initial cohesion. These findings highlight the combined influence of TPU and cooling on initial adhesion, providing an optimized strategy for high-performance bonding applications while aligning with industry demands for efficient and sustainable adhesive solutions.

Keywords: reactive polyurethane hot-melt adhesive · green strength · thermoplastic polyurethane

1 Introduction

Adhesives play a crucial role in modern industrial applications, facilitating the assembly of complex structures while offering advantages such as weight reduction, improved aesthetics and enhanced durability. Among the diverse array of adhesive technologies available, reactive polyurethane hot-melt adhesives (HMPURs) represent a highly specialized class of adhesives that have gained significant attention due to their unique combination of thermoplastic and thermosetting properties [1]. These adhesives are widely employed across multiple industrial sectors, including automotive, packaging,

L. F. M. da Silva and R. D. Adams (Eds.): AB 2025, PEM, pp. 51–70, 2026.
https://doi.org/10.1007/978-3-032-19022-2_4

woodworking, textiles and electronics, where strong bonding performance and long-term stability are essential [2, 3]. Their adaptability and growing market demand can be attributed to their high bond strength, resistance to environmental factors and ability to adhere to a wide range of substrates, including metal, plastics and composites [4]. Unlike conventional hot-melt adhesives, which rely solely on physical cooling to solidify, HMPURs undergo an additional curing process through moisture-induced polyurethane crosslinking, resulting in superior mechanical strength, enhanced chemical resistance, and long-term durability. This dual-phase bonding mechanism enables HMPURs to provide both rapid green adhesion and improved final properties, making them an attractive choice for high-performance bonding applications. Moreover, their solvent-free nature aligns with growing environmental and regulatory demands for sustainable and low-emission adhesive technologies [5]. Current environmental trends emphasize the development of bio-based alternatives and the reduction of volatile organic compounds (VOCs) in industrial adhesives, making HMPURs an attractive solution for sustainable adhesive design [6, 7].

Recent studies have demonstrated that polyurethanes formulated with polyols derived from biomass and CO_2 can offer distinct advantages in terms of adhesion strength, barrier properties and hydrolysis resistance. For instance, Chung et al. showed that the use of bio-based polycarbonate diols and polytetramethylene ether glycol (PTMEG) improved the wetting tension, tensile strength and adhesive joint performance of moisture curing PUR systems, while also contributing to hydrolysis stability and hydrogen bonding interactions highlighting the potential of renewable polyols for both performance and sustainability in PUR adhesives [8]. Similarly, Pan et al. developed isosorbide-based polyester polyols for bio-based HMPURs and showed that carbon chain length and isosorbide content strongly influenced mechanical and thermal properties, achieving adhesive strengths over 8 MPa on polycarbonate substrates, as well as good curing, thermal resistance, and tensile performance demonstrating the industrial potential of fully bio-based PUR adhesives [9].

These findings align with the growing interest improving the early-stage performance of reactive hot-melt polyurethane adhesives, particularly in terms of green strength. The reported enhancements in wetting, crystallinity and early mechanical performance associated with bio-based polyols resonate with the mechanisms underpinning green strength development [10]. However, it is important to note that these studies focused exclusively on the chemical design of the polyol structure and did not involve thermoplastic polyurethane (TPU) incorporation. In this context, the physical interactions and accelerated phase organization promoted by TPU represent a complementary strategy based on modifying the adhesive matrix through physical means. The crystallinity and rapid solidification behaviour conferred by TPU domains strengthen the adhesive during the initial bonding phase, prior to full moisture curing. Therefore, the combined use of renewable polyols and TPU-modified matrices can be considered a synergistic and still underexplored route toward optimizing green strength in HMPUR systems, while simultaneously advancing sustainability goals.

One of the main challenges in the application of HMPURs lies in improving their green strength, which refers to the initial adhesion provided by the thermoplastic matrix before moisture curing is complete. Although this early adhesion is inherent to the thermoplastic phase, it is often insufficient for demanding industrial processes where substrates must withstand mechanical stress shortly after bonding. This is particularly critical in fast-paced assembly lines or when handling heavy or geometrically complex components, where insufficient green strength may lead to premature failure or misalignment of the bonded parts. Critical in level of the green strength depends not only on the thermoplastic nature of the uncured adhesive, but also on such as the crystallinity and molecular weight of the hard segments, the viscoelastic behaviour of the matrix and the interfacial compatibility with the substrates involved [11]. To address these limitations, recent studies have explored the incorporation of organic-inorganic hybrid nanoparticles into two component polyurethane formulations, showing that such modifications can enhance green strength by improving compatibility with metallic substrates, optimizing open time, and reducing the need for additional surface treatments [12, 13].

Another promising approach to improve the performance of HMPURs is the incorporation of hydroxyl-terminated compounds, such as monomers, copolymers, and functionalized oligomers, into the adhesive formulation. These hydroxyl containing species play a crucial role in improving adhesion strength, mechanical integrity, and crosslinking control, ultimately contributing to greater durability and application efficiency across various substrates and environments [14]. Among these, the use of TPU has drawn particular attention. Owing to its partially crystalline nature and flexible chain segments, TPU can promote early crystallization and physical entanglement, reinforcing the adhesive's structural network before chemical curing takes place [15]. These physical interactions are particularly beneficial for enhancing green strength, as they provide additional cohesion during the early bonding stage, when the adhesive is still in its uncured state. Therefore, TPU not only contributes to the final properties of the adhesive once fully cured, but also significantly strengthens the joint during the critical initial minutes after application, making it a key additive in applications where fast handling and early mechanical resistance are required. For instance, Qiming Yan et al. synthesized phthalate polyester polyols (PEs) and formulated reactive polyurethane hot-melt adhesives (HMPUR). The glass transition temperature (Tg) and viscosity of the PEs decreased with increasing diol flexibility, while the initial bonding strength of the HMPUR increased with higher Tg [16]. Similarly, research on the incorporation of TPU as a green strength promoter in HMPUR has demonstrated its ability to enhance bonding performance and stability. This approach leverages the unique properties of TPUs to improve green strength while maintaining the adhesive's overall performance. TPU-modified HMPURs have also been reported to exhibit increased resistance to mechanical stress, making them suitable for dynamic applications such as footwear and flexible packaging.

However, most studies including those by Moyano et al. have been conducted on systems based on polycarbonate polyols, which exhibit different phase separation behaviour, polarity, and reactivity compared to polyester and polyether polyols or their blends. These parameters can critically affect rheology, curing behaviour, and final mechanical performance, limiting the transferability of findings to formulations typically used in industrial applications. Moreover, while Moyano et al. showed that adding 10–15 wt% TPU to polycarbonate-based HMPURs significantly enhanced green strength and reduced stringing [17], their study did not explore the effect of thermal history—particularly rapid cooling after bonding, a condition commonly encountered in automated production settings.

This gap highlights the need to investigate the influence of TPU in HMPUR systems formulated with more industrially relevant polyol blends and to understand how processing conditions such as controlled cooling affect adhesive performance.

Therefore, the present work not only investigates TPU-modified HMPURs based on mixed polyester–polyether polyols but also examines how controlled cooling influences the structural organization, viscoelastic properties, and ultimately the green strength of the adhesive. This dual approach provides a more comprehensive understanding of early-stage adhesion mechanisms and proposes a practical, synergistic strategy to enhance green strength in realistic industrial settings. Specifically, the synthesized polyurethane adhesive formulations were analysed as cured films, varying the TPU content (10 and 15 wt%), and examining their behaviour under ambient conditions and after a rapid cooling treatment. This approach enables the evaluation of both structural and processing-related effects of TPU in systems that more accurately reflect practical adhesive formulations. To our knowledge, no comprehensive study has yet addressed the combined influence of TPU loading, polyol type, and thermal history on cured HMPUR films. This work therefore contributes new insights into the formulation of high-performance, process-efficient, and potentially more sustainable reactive hot-melt adhesives.

2 Materials and Methods

2.1 Materials

Reactive polyurethane hot-melt adhesives were synthetized by using a blend of polyether and polyadipate diol polyol. For the polyether polyol, polyproplylene glycol (PPG 425, Quimidroga, S.A., Barcelona, Spain) (Mw = 425 g/mol, IOH = 250–270 mg KOH/g) was used. As polyadipate, the polyadipate of 1,4- butanediol (Hoopol F-580, Synthesia Technology, Barcelona, Spain) (Mw = 2826 g/mol). For the diisocyanate, 4,4´-diphenylmethane diisocyanate (4,4'-MDI) (Sigma Aldrich, Barcelona, Spain) was used. Pearlbond Eco 590, aromatic bio-polyurethane based on specialty polyol from renewable sources, 67% of renewable material as certified according to ASTM-D6866, was used as the TPU (Lubrizol Advanced Materials Spain S.L., Barcelona, Spain) (Tm by DSC = 64.21 °C, melt viscosity at 130 °C = 600–2000 Pa·s). Table 1 summarizes the composition of the formulated HMPUR adhesives, including the reference system and the two TPU-modified formulations. All adhesives were prepared using a 50:50 blend of polyadipate and polypropylene glycol as the polyol phase. The reference formulation (HMPUR) contains no TPU, while HMPUR-10ECO and HMPUR-15ECO incorporate 10 and 15 wt% of TPU, respectively, with corresponding adjustments in the polyol mixture to maintain the total composition.

Table 1. Composition (wt%) of the synthesized HMPUR adhesives, including the reference and TPU-containing formulations

	HMPUR	HMPUR-10ECO	HMPUR-15ECO
Polyadipate of 1,4-butanediol	50.00	50.00	50.00
Polypropylene glycol	50.00	50.00	50.00
TPU	-	17.50	26.50
4.4´-MDI	50.45	50.45	50.45

2.2 Methods

2.2.1 HMPUR Adhesive Synthesis

All adhesives were prepared using polyol mixture consisting of 50 wt% of butanediol polyadipate and 50 wt% of polypropylene glycol.

The reference adhesive was prepared by the prepolymer method with an NCO/OH ratio of 1.5 based on stoichiometric calculations. The polyols were melted and mixed at 90 °C under nitrogen atmosphere with mechanical stirring (300 rpm) for 1 h, after which 4,4'-MDI was added and mixed until the desired free isocyanate content was achieved. This content was determined by the dibutylamine method in accordance with UNE-EN 1242:2013 [18]. At this stage, the reaction was stopped, and the adhesive was stored in a hermetically sealed cartridge [19].

The adhesives HMPUR-10ECO and HMPUR-15ECO were synthesized following the same prepolymer method used for the reference formulation (Fig. 1). In these systems, TPU was incorporated into the polyol blend prior to the addition of the isocyanate. The reaction was conducted at 130 °C under mechanical stirring to ensure complete dispersion and homogeneity of the TPU within the reactive mixture.

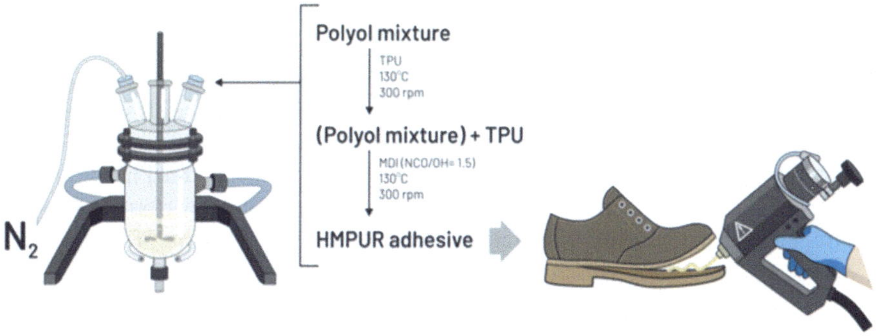

Fig. 1. HMPUR adhesives synthesis scheme

2.2.2 Characterization Techniques

To comprehensively evaluate the structural, thermal and mechanical properties of the synthesized HMPUR adhesives, a set of complementary characterization techniques was employed. These methods were selected to analyse the chemical composition, thermal transitions, viscoelastic behaviour and adhesive performance under varying conditions of TPU content and cooling rate. Two different thermal histories were applied: samples labelled as Troom were stored at 23 °C and 50% relative humidity for 7 days to allow complete ambient curing, while samples labelled as cool underwent an accelerated cooling process immediately after bonding, consisting of 10 min in a refrigeration chamber at –10 °C, followed by the same ambient curing period. These two sets of cured films were used for all subsequent analyses to assess the influence of both TPU content and thermal history on the final adhesive properties.

- *Fourier-transform infrared spectroscopy (FTIR)*: The chemical structure and functional groups of adhesives were analysed using a Cary 630 infrared spectrometer (Agilent Technologies, Santa Clara, CA, USA). The analysis was conducted using attenuated total reflection (ATR) technology, with 16 scans performed at a resolution of 4 cm^{-1}. The adhesive film was placed in direct contact with the prism for testing. Two specimens per sample were evaluated to ensure reproducibility of the results [20].
- *Thermogravimetric analysis (TGA):* The thermal behaviour of adhesives after full curing, stored for 7 days at 23 °C and 50% relative humidity, was analysed using a TGA 2 STARe System thermobalance (Mettler-Toledo AG, Schwarzenbach, Switzerland). Approximately 7–10 mg of each sample was placed on an aluminium crucible and subjected to a heating program from 30 to 600 °C at a rate of 10 °C/min under an inert nitrogen atmosphere (30 mL/min). Two specimens per sample were tested to confirm the reproducibility of the results [21].
- *Plate–plate rheology*: To assess the processability of uncured adhesives, their viscoelastic properties were evaluated using a Kinexus Pro + rheometer (Malvern Analytical, Malvern, UK) with a 25 mm diameter upper plate. The analysis involved a temperature ramp from 160 °C to 25 °C at a cooling rate of 5 °C/min, conducted at a constant frequency of 1 Hz and a controlled strain of 1%. The gap between the plates was maintained at 0.4 mm [17].
- *Differential scanning calorimetry (DSC):* To investigate the thermal transitions of fully cured adhesives, a DSC3 + Stare Systems calorimeter (Mettler-Toledo AG, Schwarzenbach, Switzerland) was used. Aluminium pans containing 9–12 mg of sample were analysed using a heating rate of 10 °C/min. The analysis consisted of three sequential thermal steps: an initial heating cycle from 30 to 120 °C (first heating), followed by a cooling cycle from 130 to -65 °C, and a final heating cycle from -65 to 135 °C (second heating). All measurements were performed under a nitrogen atmosphere with a flow rate of 50 mL/min. Two specimens of each sample were tested to confirm reproducibility [21, 22].
- *Dynamic mechanical thermal analysis (DMTA):* The viscoelastic behaviour of fully cured adhesives was assessed with DMA-Q800 equipment (TA Instruments, New Castle, DE, USA) in single-cantilever mode. Adhesive films measuring 35 mm × 13 mm × 3 mm were analysed under a nitrogen atmosphere with a heating rate of 5 °C/min, covering a temperature range from – 80 to 100 °C. The tests were performed at a constant frequency of 1 Hz, an amplitude of 20 μm and a strain of 0.5% [17].

- <u>*T-peel strength test:*</u> The adhesion properties were assessed following the EN 1392:2007 standard [23]. Footwear reference materials measuring 150 × 30 mm, including soling material (vulcanized SBR-2 rubber, Proyección Europlan XXI, Elda, Spain) and upper material (chrome-tanned split leather, Palomares Piel S.L., Elda, Spain), were joined for testing. Prior to adhesive application, both materials underwent surface preparation. The split leather was roughened at 2,800 rpm using a P100 aluminium oxide abrasive cloth (Due Emme Abrasivi, Pavia, Italy) and a roughing machine (Superlema S.A., Zaragoza, Spain). The SBR rubber was roughened and halogenated using a 2 wt% trichloroisocyanuric acid solution in ethyl acetate (Insoco, Alicante, Spain). The HMPUR was applied at 140 °C on the rubber test specimen using a heat gun. After 30 min of adhesive application, simulating standard industrial footwear production, the substrates were activated with infrared radiation at 80 °C using a CAN 02/01 temperature-controlled heater (AC&N, Elda, Spain). The materials were then joined and pressure of 1.8 bar was applied for 10 s, ensuring a consistent adhesive film thickness of 0.53 ± 0.01 mm [20, 21]. Adhesion performance was evaluated in two stages: green strength was evaluated 5 min after joint formation and after 5 min of sudden cooling (10 min at − 10 °C in a refrigerating chamber). Final adhesion was measured after storing the adhesive joints for 7 days at 23 °C and 50% relative humidity. T-peel strength tests were conducted using an Instron 34TM–10 universal testing machine (Instron Ltd., Buckinghamshire, UK) at a crosshead speed of 100 mm/min. Five specimens per adhesive were evaluated, and failure mode was assessed according to ISO 17708:2018 standard [24].

3 Results

The study evaluates the structural and chemical behaviour of the synthesized polyurethane adhesives after applying different cooling conditions and incorporating TPU into the formulation. The aim was to assess how these variables may influence both green strength and final properties.

The chemical structure of the adhesives was analysed by FTIR spectroscopy, performed on cured adhesive films subjected to two different thermal histories: storage at room temperature (roomT) and controlled cooling process at –10 °C for 10 min in a refrigeration chamber (Fig. 2). The spectra obtained for the three formulations at roomT are highly similar and exhibit the characteristic absorption bands of polyurethanes systems [21], including: the N–H stretching vibration at 3322 cm^{-1}, C–H stretching of methyl and methylene groups at 2922–2850 cm^{-1}, the carbonyl (C = O) stretching of urethane groups at 1723 cm^{-1}, and signals in the 1600–1500 cm^{-1} region associated with N–H bending and C–N stretching. Importantly, no absorption was observed at 2258 cm^{-1}, confirming the absence of residual isocyanate groups in the cured films [25].

The incorporation of TPU does not induce any modification in the chemical structure of the adhesive matrix. This is evidenced by the absence of new absorption bands or significant shifts in the FTIR spectra. The slight increase in the intensity of the C–H stretching bands cannot be ascribed to TPU, since it is also observed in the reference formulation without its incorporation. Regarding the N–H bands, although some variation in intensity is detected in the TPU-containing samples, this effect is not sufficiently significant to be considered a structural change attributable to the polymer, but rather a consequence of the formulation itself and its morphology.

Moreover, the controlled cooling treatment applied after bonding, intended to simulate sudden temperature drops during industrial processing, did not induce significant changes in the spectral profiles. The stability of the main absorption bands indicates that the polyurethane network maintains its chemical integrity regardless of the thermal history.

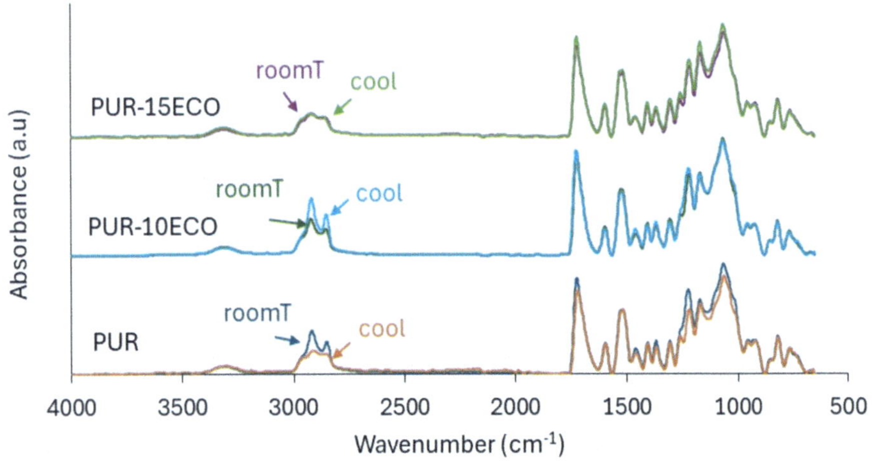

Fig. 2. Comparison of the FTIR spectra of cured HMPUR adhesive films, highlighting the effect of the initial curing conditions: room temperature (23 °C and 50% relative humidity) vs. accelerated cooling for 10 min.

The mechanical and viscoelastic behaviour of the formulations was analysed by rheology. The tests were carried out on plate-plate allowing measurement of the elastic (G′) and viscous (G″) moduli, as well as the complex viscosity (η*) over a temperature range from 20 °C to 160 °C. These parameters are essential to understand the flow characteristics, degree of structuring and solidification behaviour of the adhesives under thermal conditions relevant to their application.

Figure 3a shows the evolution of G′ and G″ with temperature for the HMPUR reference formulation and the TPU-modified HMPUR adhesives (HMPUR-10ECO and HMPUR-15ECO). As temperature increases, all formulations exhibit a progressive decrease in both moduli, consistent with the softening and gradual loss of elastic structure typical of polyurethane hot melts. However, the TPU-containing adhesives present higher values of G′ and G″ throughout the entire temperature range, indicating a stiffer and more structured network even before curing. This enhancement is attributed to the reinforcing effect of TPU, which contributes additional segmental interactions and entanglements to the matrix. The proximity of the G′ and G″ curves also suggests a more elastic like behaviour in the TPU modified systems at intermediate temperatures [26].

The complex viscosity curves presented in Figure 3b reinforce the trend observed in the shear moduli results. All adhesives' formulations show the typical exponential decrease in viscosity with temperature, consistent with thermoplastic flow behaviour. However, the incorporation of TPU leads to a systematic increase in complex viscosity (η*), with HMPUR-15ECO showing the highest viscosity values throughout the entire temperature range. This behaviour reflects a greater internal structuring and resistance to flow in the uncured state, which may contribute to improved early-stage cohesion after application. The incorporation of TPU leads to a systematic increase in complex viscosity (η*), with HMPUR-15ECO showing the highest values across the entire temperature range. This indicates a more cohesive and structured molten matrix, which resists flow and may contribute to enhanced cohesion during early stages of application. Although HMPUR-10ECO remains close to the reference formulation (HMPUR), the formulation containing 15 wt% TPU exhibits a noticeably more pronounced structuring effect, particularly at intermediate temperatures (80–120 °C) where differences between the formulations become more evident. These results support the conclusion that TPU enhances the viscoelastic properties of HMPUR adhesives not through chemical modification, but through physical reinforcement of the matrix, an effect that is especially relevant for im-proving green strength during the early bonding phase [27].

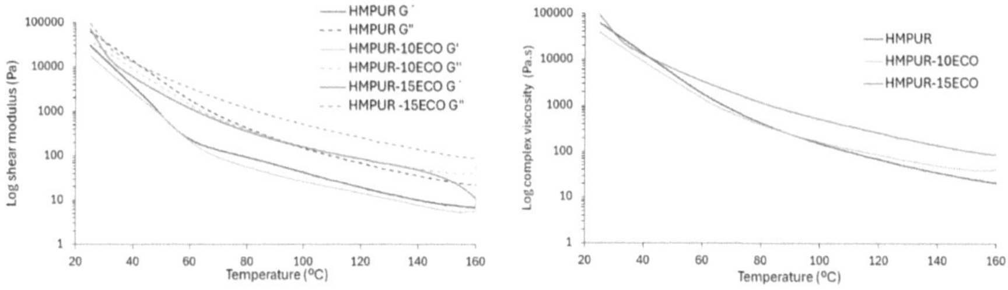

a) b)

Fig. 3. Results of plate-plate rheology experiments of synthetized HMPUR adhesives. a) Comparison of the variation of the elastic (G´) and viscous moduli (G´´) as a function of temperature and b) comparison of the variation of the complex viscosity as a function of temperature

Rheological analysis has provided a comprehensive understanding of the influence of TPU content on the viscoelastic properties of polyurethane adhesive formulations. However, a thorough evaluation of their thermal behaviour requires complementary thermogravimetric analysis (TGA), which allows the evaluation of thermal stability and decomposition under controlled conditions.

The TGA presented in Figure 4 and the data in Table 2 allow to evaluate the thermal stability and decomposition processes of the synthetized polyurethane adhesive formulations as cured films, as a function of the percentages of added TPU and according to their behaviour with and without fast cooling. The TGA (Fig. 4a) and DTGA (Fig. 4b) curves provide a detailed insight into the thermal de-composition processes of the adhesive formulations between 30 and 600 °C. This analysis highlights the differences between the standard formulations (HMPUR) and those incorporating TPU (HMPUR-10ECO and HMPUR-15ECO), as well as the effect of cooling conditions (rapid versus roomT temperature). Key temperature and weight loss values are summarized in Table 2.

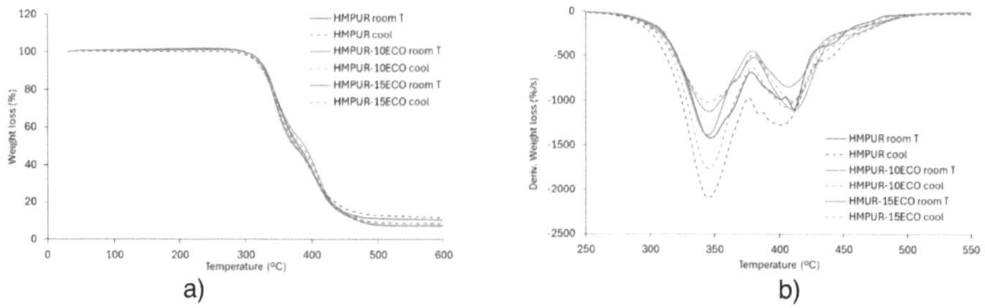

Fig. 4. Comparison of TGA (a) and DTGA (b) curves of the synthetized adhesives

Table 2. DTG and TG data of the synthetized HMPUR adhesives

Adhesive	T_1 (°C)	Wt_1 (wt%)	T_2 (°C)	Wt_2 (wt%)	Residue
HMPUR roomT	347.00	50.31	409.83	36.34	13.47
HMPUR cool	344.67	49.95	400.50	34.32	16.05
HMPUR-10ECO roomT	344.33	53.01	407.83	41.32	7.83
HMPUR-10ECO cool	346.33	49.33	403.50	40.61	8.96
HMPUR-15ECO roomT	346.50	47.57	411.50	46.55	7.65
HMPUR-15ECO cool	344.67	50.06	408.83	42.74	9.17

Firstly, the addition of TPU does not significantly alter the decomposition temperature of the hard segments (T_1), which remains in the 344–347 °C range across all systems. This suggests that the thermal stability of the highly ordered urethane domains, stabilized by hydrogen bonding, remains unaffected by either TPU incorporation or accelerated cooling. However, the impact of cooling becomes more evident in the second decomposition stage (T_2), associated with soft segment degradation: in all formulations, T_2 tends to slightly decrease under fast cooling conditions, indicating increased structural homogeneity and reduced segmental mobility within the soft domains due to kinetically trapped morphologies. Rapid solidification may restrict the full relaxation and reorganization of flexible chains, enhance the density of physical crosslinking and thus modify the degradation pattern of the polymer matrix.

Furthermore, the final residue values provide additional insight into the effect of thermal history. The HMPUR reference formulation shows an increase in char residue after accelerated cooling (from 13.47% to 16.05%), which can be attributed to tightly crosslinked network. In contrast, TPU-containing adhesives exhibit overall lower residue contents (ranging from 7.65% to 9.17%), reflecting more efficient thermal degradation and improved miscibility between hard and soft segments. This behaviour is particularly clear in the HMPUR-10ECO and HMPUR-15ECO systems, where decomposition proceeds via two well defined stages, suggesting a more homogeneous structure facilitated by the semi-crystalline nature of TPU. Taken together, these results support the conclusion that both TPU incorporation and rapid cooling synergistically enhance the thermal response of the adhesive systems. They do so not only by reinforcing the physical network but also by stabilizing the degradation pathway, which is a key consideration for high performance adhesives subjected to thermal stress during industrial processing [28, 29].

Differential scanning calorimetry (DSC) was employed to investigate the thermal transitions of the fully cured adhesive formulations, with a particular focus on the effects of TPU content and accelerated cooling treatment.

Figure 5 presents the DSC curves of the HMPURs, allowing for a deeper understanding of the crystallization and melting behaviour as a function of TPU content and post-curing thermal treatment.

Figure 5a shows the cooling cycle, while Fig. 5b corresponds to the second heating run.

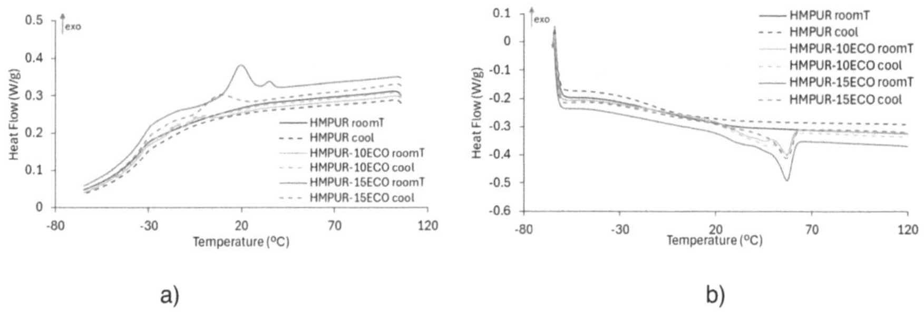

Fig. 5. Comparison of DSC curves a) cooling cycle and b) 2nd heating cycle of the HMPUR adhesives as a function of TPU content and initial curing conditions

In the cooling cycle (Fig. 5a), TPU-containing adhesives exhibit exothermic crystallization peaks absent in the unmodified HMPUR formulation, which confirms its amorphous nature and highly crosslinked structure. These crystallization exotherms are indicative of the ability of TPU domains to reorganize into semicrystalline phases upon cooling. Notably, the crystallization onset shifts toward higher temperatures as the TPU content increases, especially in HMPUR-15ECO, suggesting an enhanced tendency to crystallize due to the presence of more abundant and better organized TPU crystalline segments. This behaviour supports the idea that TPU acts as a nucleating agent within the adhesive matrix, favouring ordered packing during cooling.

However, samples subjected to rapid cooling (–10 °C for 10 min) exhibit a delay or suppression in the crystallization peaks, particularly in HMPUR-10ECO and HMPUR-15ECO, where the thermal signal becomes less intense or broader. This is attributed to kinetic trapping of the polymer chains during sudden solidification, which limits their ability to reorganize into ordered structures. This phenomenon is consistent with the formation of a denser and more physically crosslinked network, as fast cooling hinders chain mobility and disrupts the formation of well-developed crystalline domains.

In the second heating cycle (Fig. 5b), TPU-modified formulations exhibit well-defined endothermic melting transitions, absent in the reference formulation, further confirming the development of TPU crystalline phases. The position and enthalpy of these melting peaks (Table 3) are strongly influenced by TPU content: the melting enthalpy (ΔH_m) increases with higher TPU loading, reaching a maximum of 12.64 J/g for HMPUR-15ECO roomT. The peak melting temperature (T_m) also shifts slightly, with HMPUR-15ECO showing the highest T_m (~57 °C), reflecting the enhanced thermal stability of the crystalline domains.

Interestingly, accelerated cooling leads to a slight reduction in melting enthalpy in all TPU-containing systems, particularly noticeable in HMPUR-15ECO cool (9.48 J/g). This again suggests that rapid cooling partially suppresses the crystallization process, resulting in fewer or less perfect crystals being formed during the initial solidification. Nevertheless, the remaining crystalline content is sufficient to produce detectable melting peaks during reheating, indicating that TPU still retains its ability to establish ordered regions, albeit to a lesser extent.

Regarding the glass transition temperature (T_g), a systematic shift toward lower values is observed in cooled samples, as evidenced in Table 3. For example, the T_g of the reference HMPUR formulation decreases from –6.1 °C (roomT) to –12.4 °C (cool), while TPU-modified systems exhibit even lower T_g values due to the increased presence of flexible soft segments. However, in these systems, the T_g shift due to cooling is less pronounced, as the flexibility of the TPU component counteracts the stiffening effect of rapid solidification. This trend reflects the balance between two opposing mechanisms: increased physical crosslinking due to kinetic trapping and enhanced segmental mobility due to the soft nature of TPU.

Table 3. Results obtained from DSC thermograms of HMPUR adhesives 2nd heating cycle

Adhesive	T_g (°C)	T_m (°C)	ΔH_m (J/g)
HMPUR roomT	-6.13	-	-
HMPUR cool	-12.39	-	-
HMPUR-10ECO roomT	-43.35	46 and 57	11.05
HMPUR-10ECO cool	-41.65	47 and 57	10.94
HMPUR-15ECO roomT	-42.53	57	12.64
HMPUR-15ECO cool	-40.86	57	9.48

Overall, the DSC results highlight the dual role of TPU and rapid cooling in tailoring the thermal behaviour of HMPUR adhesives. TPU enhances crystallinity and contributes flexible domains, while fast cooling promotes physical crosslinking and modifies crystallization kinetics. These structural changes are essential for tuning both the green strength and long-term thermal stability of the adhesive, enabling precise control over its performance in demanding industrial applications [30, 31].

TGA results complement this interpretation of the thermal behaviour of the adhesive formulations. In the reference samples (HMPUR), accelerated cooling increases crosslinking and thermal stability, while the decrease in Tg reflects the balance between kinetic trapping and the flexibility of the soft-segment. On the other hand, in the HMPUR-10ECO and HMPUR-15ECO formulations, the simplification of the thermal profile to only two decomposition steps observed in TGA is aligned with the higher structural homogeneity detected in DSC, where TPU promotes integration between rigid and soft segments. These findings reinforce the idea that rapid cooling and the use of TPU contribute synergistically to improve both the thermal properties and the overall stability of the material.

Dynamic-mechanical thermal analysis (DMTA) was employed to evaluate the viscoelastic performance of the fully cured adhesive formulations, with a focus on the combined influence of TPU incorporation and accelerated cooling treatment [32].

Figure 6 shows the DMTA results of the fully cured HMPUR formulations, evaluating the evolution of storage modulus (E'), loss modulus (E''), and damping factor (tan δ) as a function of temperature. The key parameters derived from the analysis are summarized in Table 4 [32]. This technique complements the thermal transitions observed in DSC and the stability trends described by TGA, allowing a detailed under-standing of the viscoelastic behaviour of the adhesives as influenced by both TPU incorporation and thermal history. The storage modulus (E') showed in Fig. 6a exhibited a characteristic thermoset behaviour, with high values at sub ambient temperatures (-40 °C) corresponding to the glassy region. Upon heating, E' progressively decreased, indicating the onset of the glass transition and the transition into the rubbery regime. All formulations subjected to rapid cooling showed markedly higher E' values compared to their room temperature cured counterparts, particularly at both –40 °C and 25 °C. This enhancement suggests an increase in network rigidity induced by a denser physical crosslinking and a more constrained polymer chain arrangement. Such a response is consistent with the suppression of molecular mobility due to a kinetically frozen microstructure resulting from fast solidification. Among the TPU modified formulations, HMPUR-10ECO cool shows the most pronounced enhancement, reaching 5116 MPa at 25 °C compared to 2251 MPa for its room-temperature-cured counterpart. This significant reinforcement suggests that the semicrystalline TPU domains contribute to an internal network structure that resists deformation even before chemical curing is complete. In contrast, while HMPUR-15ECO also benefits from TPU addition, its higher soft segment content and partial phase separation lead to a somewhat lower modulus, despite still showing improvement under cooling conditions [33, 34].

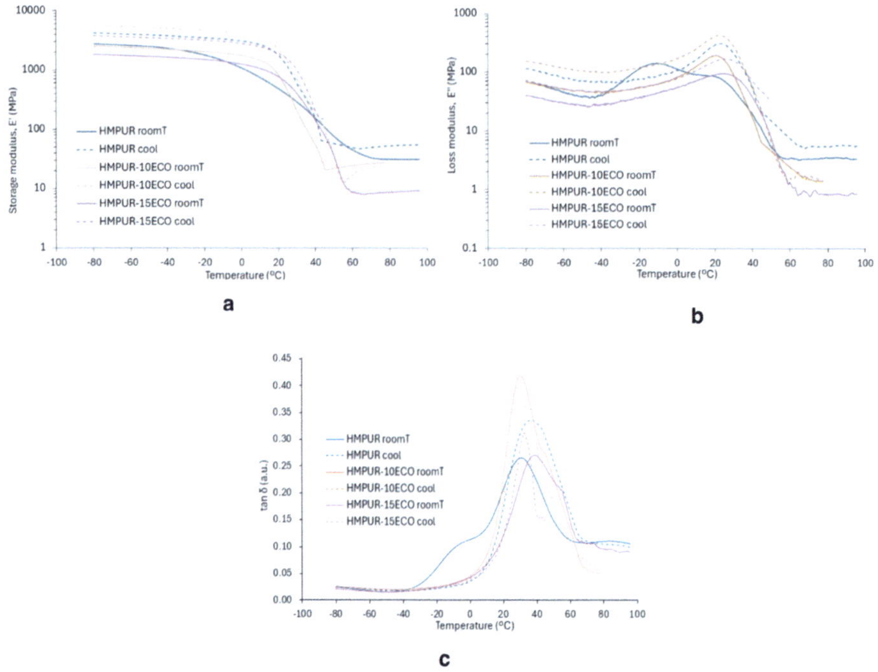

Fig. 6. Variation in the a) storage (E'), b) loss moduli (E'') and c) tan δ of the adhesives with temperature during DMTA analysis

Table 4. DMTA results. Storage modulus (E') at -40 °C and 25 °C, the Tg (°C) (calculated as the tan δ peak) and maximum value of the tan δ curve of the adhesives

Adhesive	E'(MPa) at 25 °C	E'(MPa) at –40 °C	Tg (°C)	tan δ
HMPUR roomT	2310	376	29	0.27
HMPUR cool	3759	1139	36	0.33
HMPUR-10ECO roomT	2251	456	31	0.42
HMPUR-10ECO cool	5116	1358	33	0.36
HMPUR-15ECO roomT	1633	589	39	0.27
HMPUR-15ECO cool	2574	956	33	0.23

The loss modulus (E'') curves presented in Fig. 6b reflect the viscous response of the adhesive systems and highlight the segmental relaxation processes occurring within the polymer matrix. All formulations display a prominent α-relaxation peak, associated with the glass transition of the soft segments. A systematic shift of this peak toward higher temperatures is observed in all samples subjected to accelerated cooling, indicating a restriction in chain mobility caused by increased physical crosslinking and reduced free volume. The position and shape of the E" peak are highly sensitive to both the thermal history and the TPU content of the formulation.

The incorporation of TPU modifies both the amplitude and sharpness of the relaxation peak. In HMPUR-10ECO, the peak appears sharper and more intense, suggesting a more defined glass transition, likely due to the contribution of semicrystalline TPU domains that reinforce the network. In contrast, HMPUR-15ECO exhibits a broader and less intense peak, indicative of enhanced energy dissipation, which can be attributed to a higher proportion of soft segments and a partial suppression of TPU crystallinity. In all cases, the effect of rapid cooling narrows the relaxation window and shifts the glass transition toward higher temperatures, particularly in the reference HMPUR, whose initially amorphous structure allows for greater kinetic trapping during cooling.

The tan δ curves shown in Fig. 6c, defined as the ratio of E" to E', provide additional insights into the damping capacity and energy dissipation of the adhesives. Each formulation shows a single dominant tan δ peak, corresponding to the glass transition. As with E", the tan δ peak systematically shifts to higher temperatures after accelerated cooling, while its intensity decreases. This trend suggests a transition toward more elastic behaviour and reduced damping, consistent with a denser and more physically crosslinked network. In this case, the glass transition temperature (Tg) was determined from the maximum of the tan δ peak. The influence of cooling on Tg is clearly formulation-dependent: in the reference HMPUR, Tg increases after accelerated cooling, reflecting restricted chain mobility and a denser physical network; in HMPUR-10ECO, Tg remains practically unchanged, as the flexibility introduced by TPU counterbalances the stiffening effect; whereas in HMPUR-15ECO, Tg decreases, evidencing the predominance of soft segments and the reduction of crystallinity.

Interestingly, the highest tan δ values are found in HMPUR-10ECO at room temperature, reflecting optimal energy dissipation and chain mobility prior to solidification. Upon cooling, the tan δ peak in this formulation is significantly reduced, indicating a stiffening of the matrix and a reduction in molecular motion due to the frozen morphology. In HMPUR-15ECO, the broader and flatter tan δ peaks suggest a more heterogeneous distribution of relaxation modes, likely caused by microphase separation and lower compatibility between hard and soft segments.

Adhesion tests were conducted to evaluate both the green and final strength of the newly developed formulations. Green strength was assessed 5 min after bond formation or 5 min following the accelerated cooling treatment. Figure 7 summarises the T-peel strength results for split-leather/adhesive/SBR joints, revealing no green strength for the joints prepared with the reference formulation at room temperature nor after fast cooling in the reference formulation. However, adhesion improves with the incorporation of TPU, reaching 0.7 N/mm for HMPUR-10ECO roomT and 0.9 N/mm for HMPUR-15ECO roomT, with a more pronounced enhancement observed upon applying accelerated cooling after bonding (2.4 N/mm for HMPUR-10ECO cool and 2.7 N/mm for HMPUR-15ECO cool). HMPURs are applied at a relatively low temperature (90–140 °C) on substrates. At such temperatures they flow well and can adequately wet the surface of the adhering material. After that, they undergo a two-phase curing process. In the first phase, also called physical crosslinking, green strength develops due to the cooling process, together with the crystallization of the soft segments that can be found in the polyol's component. The presence of the TPU-ECO with high thermoplasticity improves green strength by accelerating the crystallization of the adhesive. The accelerated crystallization by fast cooling further accentuates this effect, which is not present in the reference formulation [15, 35], confirming the synergistic effect of TPU and fast cooling on the increased mechanical and green adhesion properties.

In TPU-containing formulations, the improvement in green adhesion is closely related to the intrinsic properties of this polymer. Its semicrystalline nature and high thermoplasticity promote rapid physical crosslinking through the crystallization of soft segments during cooling. This facilitates the formation of a transient network that strengthens adhesive cohesion within minutes of application. Additionally, TPU acts as a nucleating agent, accelerating crystallization and enabling the adhesive to transition more quickly from the molten state to a solid structure, even before chemical curing begins.

Accelerated cooling further enhances this effect: the rapid temperature drop kinetically traps the crystalline TPU domains, increasing rigidity and reducing molecular mobility. This dense packing of polymer chains leads to an immediate rise in cohesive strength, which is not observed in the reference formulation, as it lacks crystallizable segments and remains soft and mechanically weak until chemical curing progresses. This behaviour is clearly reflected in Fig. 7, where TPU-containing formulations subjected to accelerated cooling exhibit a significant improvement in green adhesion, as compared to their non-cooled counterparts.

However, this early crystallization may also introduce limitations. TPU increases melt viscosity and promotes microphase separation between hard and soft domains, resulting in premature rigidification of the adhesive matrix. These morphological changes, confirmed by DSC and DMTA analyses, hinder the mobility of isocyanate groups and their diffusion toward the substrate interface, reducing the number of covalent bonds that can form during chemical curing. Moreover, semicrystalline domains may act as physical barriers, disrupting the continuity of the reactive network. Although TPU and accelerated cooling enhance initial cohesion, they may compromise final adhesion if the adhesive does not achieve optimal wetting and interfacial diffusion.

The second phase of HMPUR curing consists of chemical crosslinking, in which isocyanate groups react with ambient moisture and/or the moisture from the substrates. As a result, the molecular weight increases and a fully reacted polymer is obtained, where urethane and urea groups alternate, forming an irreversible thermoset structure that provides the final adhesive strength. In our study, final adhesion was evaluated after 7 days of curing. Although all samples achieved high final peel strength values ($>$6.4 N/mm), a reduction in T-peel strength was observed in cooled samples compared to those cured at room temperature.

This phenomenon may be attributed to the adhesive's reduced fluidity during the bonding process. In samples cured at room temperature, the adhesive remains fluid for a longer time, allowing it to penetrate the microroughness of the SBR substrate and react with chlorine species from the halogenation process. In contrast, in samples subjected to accelerated cooling, the adhesive undergoes a rapid increase in viscosity, limiting migration and diffusion. Consequently, fewer interfacial linkages form with the rubber material.

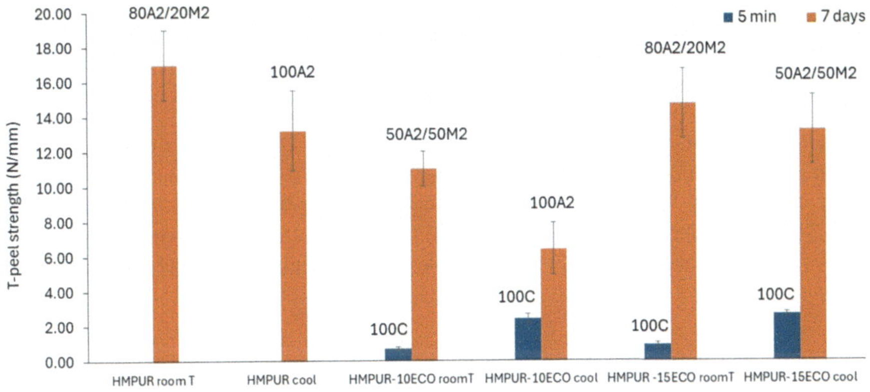

Fig. 7. T-peel strength of the split-leather/adhesive/SBR 5 min and 7 days after bonding. Failure mode (ISO 17708:2003E): A2 separation of the adhesive film from sole material (SBR); M2 partial or complete breaking of sole material; C defective cohesion of the adhesive

4 Conclusions

This study has systematically evaluated the impact of thermoplastic polyurethane (TPU) incorporation and controlled cooling on the structural, thermal and mechanical performance of polyester–polyether based reactive polyurethane hot-melt adhesives (HMPURs). The results demonstrate that both TPU content and accelerated cooling play critical roles in enhancing the green strength and modifying the thermal and viscoelastic behaviour of the adhesives.

Thermal analysis through TGA and DSC highlighted that rapid cooling contributes to a higher degree of molecular organization, reflected in altered decomposition profiles and increased glass transition temperatures (T_g), except for the reference formulation which shows a decrease in Tg when accelerated cooling is applied. TPU-modified formulations exhibited more homogeneous thermal behaviour. Additionally, rapid cooling delayed the crystallization process, especially in TPU-containing adhesives, due to restricted chain mobility caused by enhanced crosslinking.

Dynamic mechanical thermal analysis (DMTA) confirmed these findings, showing increased storage and loss moduli, as well as higher T_g values for cooled samples. The highest stiffness was observed in the HMPUR-10ECO formulation, while HMPUR-15ECO maintained a balance be-tween flexibility and mechanical integrity, suggesting that TPU content can be tuned to achieve specific mechanical profiles depending on the application requirements.

Adhesion testing demonstrated the synergistic effect of TPU and accelerated cooling on green strength development. The TPU-modified adhesives exhibited significantly improved green strength compared to the reference formulation, with further enhancement achieved through rapid cooling. Although final adhesion strength remained high across all formulations, some reduction was observed in cooled samples, likely due to limited adhesive diffusion and interfacial interaction with the substrate.

Overall, the incorporation of TPU, combined with controlled cooling, offers a practical and effective approach to improving the early-stage performance of reactive polyurethane hot-melt adhesives without compromising long-term adhesion strength.

These results underscore the potential of TPU-modified HMPURs for applications requiring im-proved green strength and enhanced thermal stability, paving the way for further optimization in adhesive formulations.

Author Contributions. Conceptualization was carried out by M.D.R.-S, M.P.C-B and A.M.-V. Methodology; validation, investigation and original draft preparation were conducted by A.M.-V, C.H.-F and M.P.C-B. Writing—review and editing- was performed by A.M.-V, C.H.-F, M.P.C-B. and M.D.R.-S. Supervision was provided by E.O.-C and M.D.R.-S. Funding acquisition was managed by E.O.-C and M.D.R.-S. All authors have read and agreed to the published version of the manuscript.

Funding. This research was funded by the Valencian Institute for Business Competitiveness (IVACE) and the European Regional Development Fund (ERDF), grant number IMDEEA/2024/43, CIRCADH II.

Institutional Review Board Statement: Not applicable.

Data Availability Statement:. The original contributions presented in the study are included in the article, further inquiries can be directed to the corresponding author.

Conflicts of Interest:. Authors, Carbonell-Blasco María Pilar, Moyano-Vallejo Alejandra, Hernández-Fernández Carlota, Romero-Sánchez María Dolores and Orgilés-Calpena Elena were employed by the company INESCOP Footwear Technology Centre. The authors declare no conflicts of interest. The funders had no role in the design of the study; in the collection, analyses, or interpretation of the data; in the writing of the manuscript; or in the decision to publish the results.

References

1. Gharde, S., Sharma, G., Kandasubramanian, B.: Hot-Melt Adhesives: Fundamentals, Formulations, and Applications: A Critical Review. In: Mittal, K.L. (ed.) Progress in Adhesion and Adhesives, pp. 1–28. Wiley (2021)
2. Bautista-Anguís, D., et al.: Synthesis and characterization of rebondable polyurethane adhesives relying on thermo-activated Transcarbamoylation. Polymers **16**, 2799 (2024). https://doi.org/10.3390/polym16192799
3. Tang, Q., He, J., Yang, R., Ai, Q.: Study of the synthesis and bonding properties of reactive hot-melt polyurethane adhesive. J. Appl. Polym. Sci. **128**, 2152–2161 (2013). https://doi.org/10.1002/app.38415
4. Pizzi, A., Mittal, K.L.: Handbook of Adhesive Technology. CRC Press (2017)
5. Josyula, S.P., Diebels, S.: Numerical Investigation of Fracture Behaviour of Polyurethane Adhesives under the Influence of Moisture. Polymers **16**, 2676 (2024). https://doi.org/10.3390/polym16182676
6. Tian, S.: Recent advances in functional polyurethane and its application in leather manufacture: a review. Polymers **12**, 1996 (2020). https://doi.org/10.3390/polym12091996
7. Arias, A., et al.: Recent developments in bio-based adhesives from renewable natural resources. J. Clean. Prod. **314**, 127892 (2021). https://doi.org/10.1016/j.jclepro.2021.127892
8. Chung, C.-H., Shih, W.-C., Chiu, W.-M.: Synthesis, characterization and properties of biomass and carbon dioxide derived polyurethane reactive hot-melt adhesives. E-Polym. **19**, 535–544 (2019). https://doi.org/10.1515/epoly-2019-0057

9. Pan, X., Tian, Y., Li, J., Tan, Q., Ren, J.: Bio-based polyurethane reactive hot-melt adhesives derived from isosorbide-based polyester polyols with different carbon chain lengths. Chem. Eng. Sci. **264**, 118152 (2022). https://doi.org/10.1016/j.ces.2022.118152
10. Arán-Ais, F., Ruzafa-Silvestre, C., Carbonell-Blasco, M., Pérez-Limiñana, M., Orgilés-Calpena, E.: Sustainable adhesives and adhesion processes for the footwear industry. Proc. Inst. Mech. Eng. Part C J. Mech. Eng. Sci. **235**, 585–596 (2021). https://doi.org/10.1177/0954406220957706
11. Chen, Y., Hu, J., Wu, X., Duan, S., Wang, H., Ma, T.: Synthesis and Evaluation of Polyurethane as Waterproof Adhesion Layer for Steel Bridge Deck. Polymers **16**, 3140 (2024). https://doi.org/10.3390/polym16223140
12. Duong, T., Le, M., Lee, C., Kim, J.: Organic-Inorganic Hybrid Nanoparticles for Enhancing Adhesion of 2K Polyurethane to Steel and Their Performance Optimization Using Response Surface Methodology. Polymers **16**, 2816 (2024). https://doi.org/10.3390/polym16192816
13. Leitsch, E.K., Heath, W.H., Torkelson, J.M.: Polyurethane/polyhydroxyurethane hybrid polymers and their applications as adhesive bonding agents. Int. J. Adhes. Adhes. **64**, 1–8 (2016). https://doi.org/10.1016/j.ijadhadh.2015.09.001
14. Tahir, N.A., Ahmad Zubir, S.: Investigating the Impact of Systematic Hard Segment Modulation on Performance of Hydroxyl Terminated Polybutadiene (HTPB) Thermoset Polyurethane Elastomers. J. Appl. Polym. Sci. **142**, e56510 (2025). https://doi.org/10.1002/app.56510
15. INNOVATIVE PEARLBOND AND PEARLSTICK TPU SOLUTIONS FOR ADHESIVES. https://espanol.lubrizol.com/solutions/technologies/tpu/adhesives/hot-melt-adhesives
16. Yan, Q., Lin, Z., Lin, X., Fang, C.: Preparation and characterization of high initial bonding strength reactive hot melt polyurethane adhesive derived from phthalate polyester polyols. J. Adhes. Sci. Technol. **32**, 2282–2297 (2018). https://doi.org/10.1080/01694243.2018.1471841
17. Moyano-Vallejo, A., Carbonell-Blasco, M.P., Hernández-Fernández, C., Arán-Aís, F., Romero-Sánchez, M.D., Orgilés-Calpena, E.: Enhanced Green Strength in a Polycarbonate Polyol-Based Reactive Polyurethane Hot-Melt Adhesive. Polymers **16**, 3356 (2024). https://doi.org/10.3390/polym16233356
18. UNE-EN 1242-2013 Adhesives - Determination of isocyanate content
19. Orgilés-Calpena, E., Arán-Aís, F., Torró-Palau, A.M., Montiel-Parreño, E., Orgilés-Barceló, C.: Synthesis of polyurethanes from CO2-based polyols: A challenge for sustainable adhesives. Int. J. Adhes. Adhes. **67**, 63–68 (2016). https://doi.org/10.1016/j.ijadhadh.2015.12.027
20. Carbonell-Blasco, M.P., Pérez-Limiñana, M.A., Ruzafa-Silvestre, C., Arán-Ais, F., Orgilés-Calpena, E.: Influence of biobased polyol type on the properties of polyurethane hotmelt adhesives for footwear joints. Appl. Adhes. Sci. **9**, 8 (2021). https://doi.org/10.1186/s40563-021-00138-7
21. Carbonell-Blasco, M.P., Ruzafa-Silvestre, C., Mateu-Romero, B., Orgilés-Calpena, E., Arán-Ais, F.: Cleaner technologies to minimise the environmental footprint of the footwear bonding process. Proc. Inst. Mech. Eng. Part C J. Mech. Eng. Sci. 09544062221140053 (2022). https://doi.org/10.1177/09544062221140053
22. Jin, X., Guo, N., You, Z., Tan, Y.: Design and Performance of Polyurethane Elastomers Composed with Different Soft Segments. Materials. **13**, 4991 (2020). https://doi.org/10.3390/ma13214991
23. EN 1392:2007 standard: Adhesives for Leather and Footwear Materials-Solvent-Based and Dispersion Adhesives-Testing of Bond Strength under Specified Conditions, https://www.une.org/encuentra-tu-norma/busca-tu-norma/norma?c=N0038449
24. ISO 17708:2018 Footwear — Test methods for whole shoe — Upper sole adhesion, https://www.iso.org/standard/68281.html

25. Jang, N.G., Linh, T.Q., Toan, M., Kwon, K., Shin, S.: Enhancing the Mechanical and Adhesive Properties of Polyurethane Adhesives with Propylene Oxide-Modified Ethylenediamine (PPO-EDA). Polymers **17**, 231 (2025). https://doi.org/10.3390/polym17020231
26. Fuensanta, M., Martín-Martínez, J.M.: Structural and Viscoelastic Properties of Thermoplastic Polyurethanes Containing Mixed Soft Segments with Potential Application as Pressure Sensitive Adhesives. Polymers **13**, 3097 (2021). https://doi.org/10.3390/polym13183097
27. Fuensanta, M., Vallino-Moyano, M.A., Martín-Martínez, J.M.: Balanced Viscoelastic Properties of Pressure Sensitive Adhesives Made with Thermoplastic Polyurethanes Blends. Polymers **11**, 1608 (2019). https://doi.org/10.3390/polym11101608
28. Lin, S.-Y., Ahmad, N., Jeffrey Kuo, C.-F.: Study of Synthesis of Dual-Curing Thermoplastic Polyurethane Hot-Melt Adhesive and Optimization by Using Gray Relational Analysis to Apply in Fabric Industry to Solve Seamless Bonding Issues. Polymers **16**, 467 (2024). https://doi.org/10.3390/polym16040467
29. Maminski, M., Jaskolowski, W., Parzuchowski, P.: Thermal stability of novel polyurethane adhesives investigated by TGA. Mediterr. J. Chem. **3**, 877–882 (2014). https://doi.org/10.13171/mjc.3.3.2014.17.05.13
30. Yang, J., Pan, H., Li, X., Sun, S., Zhang, H., Dong, L.: A study on the mechanical, thermal properties and crystallization behavior of poly(lactic acid)/thermoplastic poly(propylene carbonate) polyurethane blends. RSC Adv. **7**, 46183–46194 (2017). https://doi.org/10.1039/C7RA07424G
31. Batista, N.L., Olivier, P., Bernhart, G., Rezende, M.C., Botelho, E.C.: Correlation between degree of crystallinity, morphology and mechanical properties of PPS/carbon fiber laminates. Mater. Res. **19**, 195–201 (2016). https://doi.org/10.1590/1980-5373-MR-2015-0453
32. Introduction to Dynamic Mechanical Analysis and Its Application to Testing of Polymer Solids Available online:, https://www.tainstruments.com/pdf/literature/TA441.pdf
33. Bajsić, E.G., Rek, V.: Dynamic mechanical study of thermoplastic polyurethane/ polypropylene blends. E-Polym. 4 (2004). https://doi.org/10.1515/epoly.2004.4.1.837
34. Xie, W., Yan, Q., Fu, H.: Study on novel rosin-based polyurethane reactive hot melt adhesive. Polym. Adv. Technol. **32**, 4415–4423 (2021). https://doi.org/10.1002/pat.5443
35. Bockorny, G.D.A., Preuss, N.L., Gomes, C.R.I., Domingues Junior, N.S., Forte, M.M.D.C.: Bonding properties and crystallization kinetics of thermoplastic polyurethane adhesive. Matér. Rio Jan. 23 (2018). https://doi.org/10.1590/s1517-707620180004.0581

Adhesion and surface treatments

Considerations on the Criterion of Superhardness of Tool Materials in the Aspect of Coating Adhesion to Cutting Blades

Maciej Jan Kupczyk(✉)

Poznan University of Technology, Institute of Mechanical Technology, Ul. Piotrowo 3, 60-965 Poznan, Poland
maciej.kupczyk@put.poznan.pl

Abstract. The article characterizes the previous views and presents a new approach to the assessment of superhardness of tool materials - that is, a criterion that allows distinguishing hard from superhard tool materials. Due to the fact that there are large discrepancies in this respect, a new criterion of superhardness was proposed, more appropriate to modern tool materials for cutting blades, which, due to their significant resistance to abrasive wear, provide high machining efficiency even for difficult-to-cut materials. The criterion of superhardness was determined on the basis of tests that allowed observing the size of plastic deformation under the influence of a loaded diamond stylus. The adopted criterion of superhardness of materials is a practical guide for manufacturers of tools with anti-wear coatings in the aspect of their adhesion to the substrate. As stated, coatings characterized by very limited deformation capacity and high brittleness (superhard) constitute a serious limitation to their effective use in the form of thin coatings directly deposited on relatively soft substrates, such as high-speed steels. In order to systematize superhard materials, a pyramid of hardness of these materials was proposed, built at the base on compounds of nonmetals from the third, fourth, fifth and sixth groups (IIIA to VIA) of the second period of the periodic table of elements.

Keyword: New pyramid of superhard materials · Stishovite · Cutting tools

1 Introduction

1.1 Outline of the Problem

Progress in the field of materials for cutting blades is closely related to the emergence of new construction materials with better utility properties. The use of new construction materials is often associated with a decrease in machining efficiency due to their difficult cutting properties. For this reason, attempts are being made to increase the cutting efficiency of difficult-to-cut materials, among others, by using blades made of materials that are more resistant to abrasive wear. Abrasive wear of cutting blades can be limited mainly by manufacturing them from very hard materials. The works (Burakowski 2004;

L. F. M. da Silva and R. D. Adams (Eds.): AB 2025, PEM, pp. 73–85, 2026.
https://doi.org/10.1007/978-3-032-19022-2_5

Khrushhov et al. 1970; Kupczyk 2015) indicate a linear increase in relative resistance to abrasive wear with increasing material hardness. For this reason, superhard materials for cutting tool blades are becoming increasingly popular.

In practice, two approaches are used to increase the resistance of cutting blades to abrasive wear. The first one involves using very expensive bulk superhard materials. The second one involves applying thin, micrometrically thick superhard coatings to cutting blades made of traditional tool materials. As stated (Ciszewski et al. 1993), superhard material coatings are characterized by a very limited ability to plastically deform and high brittleness (superhard), which is a serious limitation of their effective use in the form of thin coatings directly deposited on relatively soft substrates, such as high-speed steels. This article synthetically characterizes the previous views on the criterion of superhardness of materials. Based on the analysis of the literature and own research, a new criterion of superhardness was proposed, which is more appropriate for modern tool materials. A hierarchical hardness system of superhard materials (the so-called new pyramid of superhard materials) has been proposed, built on nonmetals from groups IIIA to VIA of the second period of the periodic table of elements.

1.2 Current Views on Superhardness of Materials

The literature presents different approaches to the issue of dividing materials into hard and superhard. Dobrzański (2002) assumed that superhard materials are those whose hardness is at least 9 on the 10-point Mohs scale, i.e. about 20 GPa.

Similarly, Ciszewski, Przetakiewicz and Trębiński (Ciszewski et al. 1993; Trębiński 1992) define superhard materials as those whose hardness is greater than the hardness of natural corundum (Al_2O_3), i.e. with a hardness of at least 20 GPa.

The identical criterion for superhard materials was adopted by Novikov (Novikov 1986; Olszyna 2001) (at least 20 GPa), who, based on this assumption, built a pyramid of hardness of superhard materials, basing the pyramid on corundum (Al_2O_3). In the proposed pyramid, compounds considered to be superhard are arranged in the following order:

1) Al – 3rd period, group IIIA (chemical compound: Al_2O_3);
2) Si – 3rd period, group IVA (chemical compounds: SiC, B_4Si, Si_3N_4);
3) B – 2nd period, group IIIA (chemical compounds: BP, B_4Si, B_4C, BN);
4) C – 2nd period, group IVA (chemical compounds: SiC, B_4C, C- diamond);
5) N – 2nd period, group VA (chemical compounds: Si_3N_4, BN);
6) P – 3rd period, group VA (chemical compound: BP);
7) O – 2nd period, group VIA (chemical compound: Al_2O_3).

The hardness pyramid shows that superhard materials are compounds of nonmetals from the third, fourth, fifth and sixth groups of the second and third periods of the periodic table of elements (i.e. boron, carbon, nitrogen, silicon, phosphorus and oxygen), e.g. c-BN, B_4C, BP, Si_3N_4.

2 Methodology of Building a New Arrangement of Superhard Materials

Analyzing the assumptions given in the above works, two fundamental inconveniences can be indicated, namely:

they assume that a superhard material is a material with a hardness of only 20 GPa,
the elements included in the superhard materials in the Novikov pyramid have not been arranged at the base of the pyramid in the order of the groups of the periodic table of elements.

Taking into account the currently used tool materials, both in the form of bulk and coating materials, and the author's own previous research (Kupczyk 2015), the author of the article believes that this criterion is not appropriate, taking into account, among others, the plastic deformation of the material under the influence of the applied pressure, as well as the related aspect of the practical application of the materials. Therefore, this criterion was clarified by assuming that a superhard material is one that does not undergo plastic deformation, i.e. it was assumed that for tool materials with hardness above a certain limit, no visible plastic deformation will be observed under the pressure of a Rockwell diamond indenter. This is of significant importance in practice, especially in relation to anti-wear coatings deposited on relatively soft substrates, e.g. high-speed steel, which deforms significantly under the influence of high unit pressures (Kupczyk 2015). The observed mechanism of the lack of possibility of super hard coating deformation together with a much softer substrate weakens the adhesion of the coating to the substrate (Kupczyk 2015). In connection with the above, studies were carried out on the effect of pressure exerted by a Rockwell diamond indenter during the scratch test on the occurrence of plastic deformations of basic tool materials commonly used for cutting blades. These include:

HS6–5-2 high-speed steel with a hardness of 9 GPa,
Al_2O_3 (Sumitomo W80) oxide (white) ceramics with a hardness of 22–23 GPa,
Al_2O_3 + TiC mixed ceramics (Sumitomo NB90S) with a hardness of 30 GPa,
cubic boron nitride (WURBON BN5 – SPK Feldmühle) with a hardness of 33–35 GPa.

3 Results

The types and individual phases of deformation (damage) of the above-mentioned tool materials under the influence of a loaded Rockwell diamond indenter are illustrated in Figs. 1, 2, 3, 4, 5, 6, 7 and 8.

Diamond stylus load a) 20 N, b) 40 N, c) 70 N, d) 130 N.

Diamond stylus load a) 15 N, b) 20 N, c) 160 N.

Diamond stylus load a) 10 N, b) 50 N, c) 90 N.

Diamond stylus load a) 10 N, b) 20 N, c) 40 N.

Based on observations, it was found that only the substrate made of cubic boron nitride with a hardness of 33–35 GPa (WURBON in BN5 – SPK Feldmühle) does not undergo visible plastic deformation. Even in the case of mixed ceramics (Al_2O_3 + TiC)

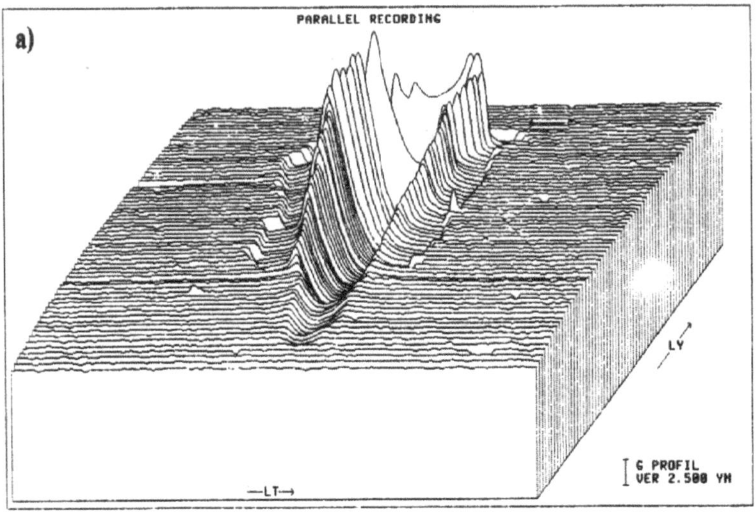

Fig. 1. Spatial image of the scratched sample made of high-speed steel HS6–5-2 with increasing load value of the diamond stylus with visible significant accumulation of material on the edges of the scratch

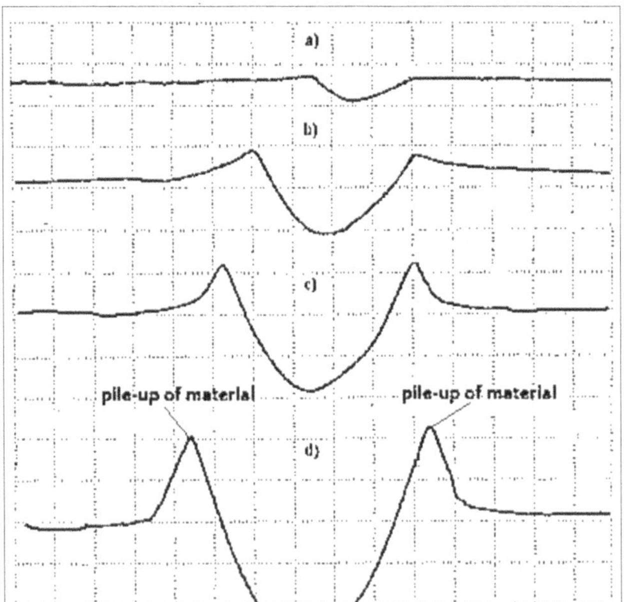

Fig. 2. Individual phases of plastic deformation of HS6-5-2 high-speed steel with hardness of 9 GPa

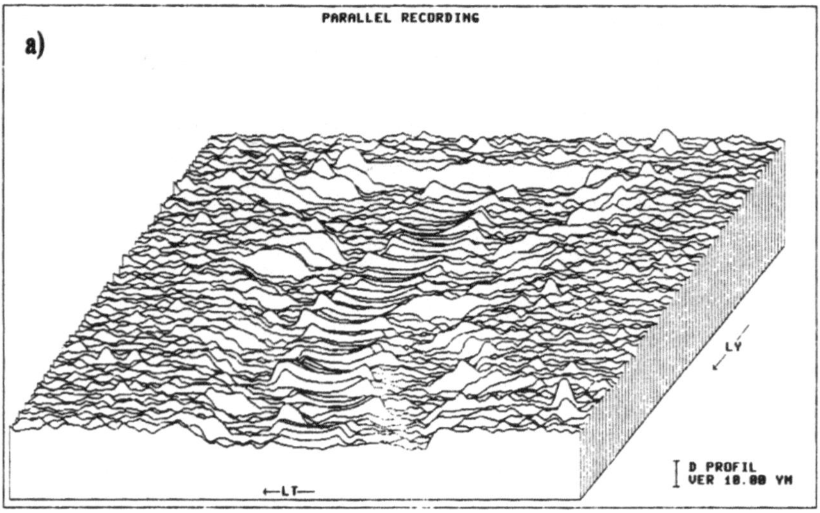

Fig. 3. Spatial image of a scratched sample made of Al_2O_3 tool ceramics (Sumitomo W80) with increasing load value of the diamond stylus with visible pile-up of material on the scratch edges

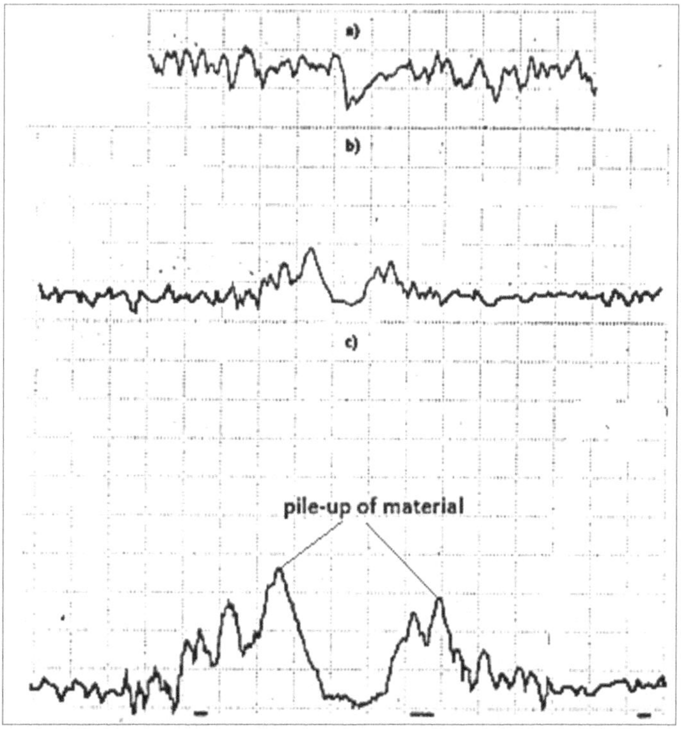

Fig. 4. Individual phases of plastic deformation of Al_2O_3 tool ceramics (Sumitomo W80) with hardness of 23–24 GPa with visible pile-up of substrate material on the scratch edges

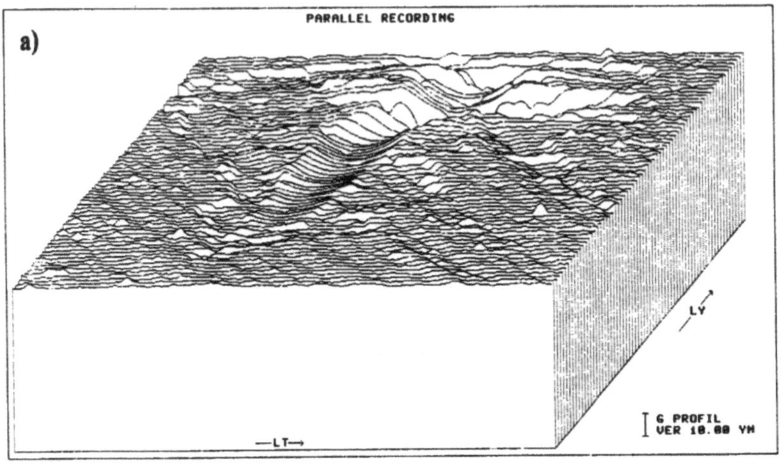

Fig. 5. Spatial image of a scratched sample made of Al_2O_3 + TiC tool ceramics (Sumitomo NB90S) with increasing load of the diamond stylus with visible pile-up of material on the scratch edges

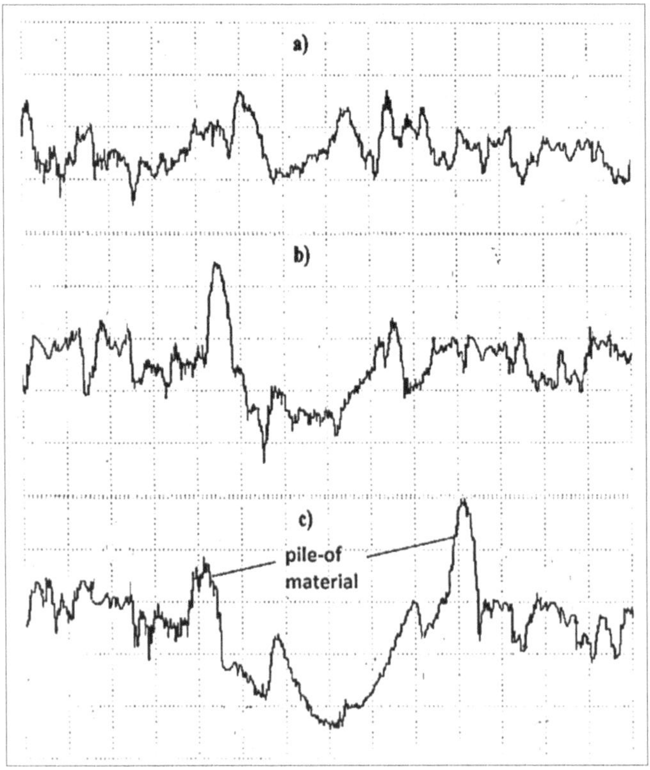

Fig. 6. Individual phases of plastic deformation of Al_2O_3 + TiC tool ceramics (Sumitomo NB90S) with hardness of 30 GPa with visible pile-up of tool material on the scratch edges

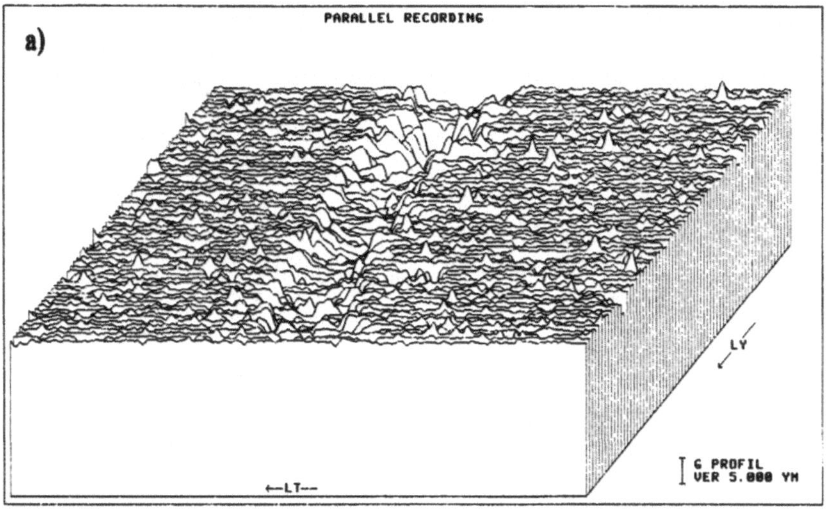

Fig. 7. Spatial image of a scratched boron nitride sample (WURBON BN5 – SPK Feldmühle) with a hardness of 33–35 GPa with an increasing load value of the diamond stylus, without visible material pile-up on the scratch edges

with a hardness of 30 GPa, it is observed. Therefore, a new criterion for the superhardness of materials of at least 33 GPa was adopted.

Due to the new criterion adopted, the Novikov hardness pyramid should be modified. The concept of building a hierarchical system of hardness of superhard materials according to the new principles is presented in Fig. 9. It was assumed that superhard materials are compounds of nonmetals from the third, fourth, fifth and sixth group of the second period of the periodic table of elements, i.e.:

group III A – B (boron),
group IV A – C (carbon),
group V A – N (nitrogen),
group VI A – O (oxygen),

containing borides, silicides, carbides, nitrides and oxides.

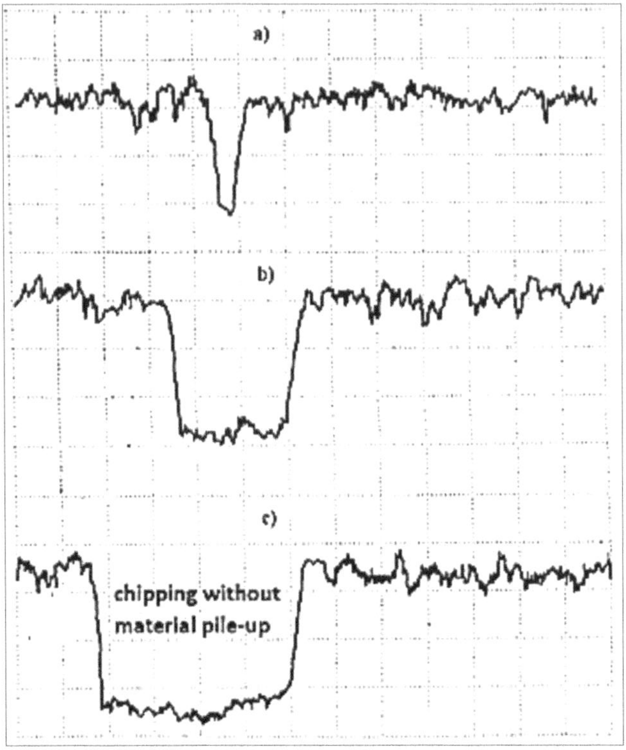

Fig. 8. Successive stages of material loss from boron nitride (WURBON w BN5 – SPK Feldmühle) with hardness of 33–35 GPa without visible plastic deformation at the scratch edges

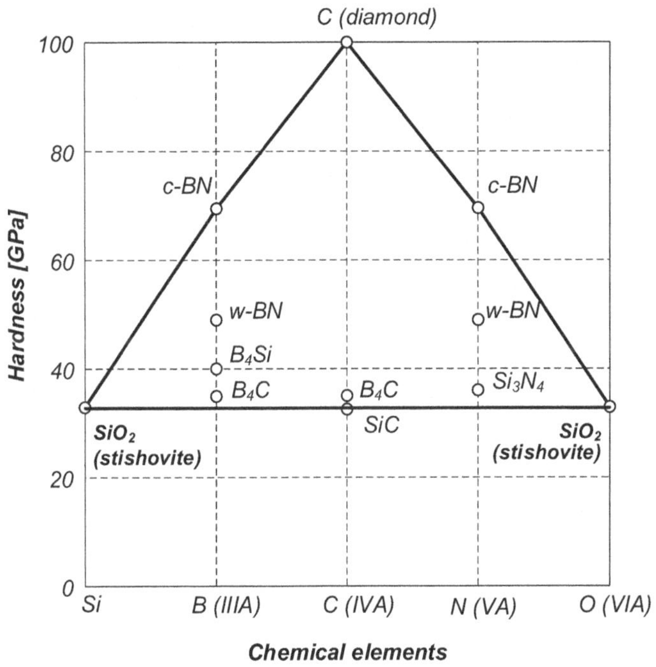

Fig. 9. New hardness pyramid of super hard materials

4 Discussion

In the presented system, these elements are arranged on the abscissa axis in the above-mentioned order (Fig. 9), which is consistent with their occurrence in the subsequent groups of the periodic table of elements (Kupczyk 2015). The new model in the base (extreme positions) instead of corundum (Al_2O_3) has a high-pressure variety of silicon oxide SiO_2 with a hardness of 33–35 GPa, obtained for the first time by Stishov and Popov (Rediel 1994; Stishov et al. 1961)). High-pressure silicon oxide is known as stishovite and is the only oxide that meets the superhardness criterion (Badzian 1988; Kupczyk 2015). The great importance of stishovite can be evidenced by the currently growing interest in this material (Nishiyama et al. 2012). The rest of the pyramid base is made up of SiC (33.4–35 GPa), Si_3N_4 (33–35 GPa), B_4C (33–40 GPa) (Dobrzański 2002; Holleck et al.; Kupczyk 2015; Olszyna 2001; Vereshhaka et al. 1986; Wysiecki 1997). In light of the above considerations, the hierarchical hardness system of materials meeting the superhardness criterion (min 33 GPa) includes (Holleck 1984; Karczemska et al. 2008; Kupczyk 2015; Leger et al.; Novikov 1986; Vereshhaka et al. 1986; Wysiecki 1997):

superhard nitrides – hexagonal boron nitride w-BN with a wurtzite structure and hardness of 47–50 GPa and c-BN with hardness of 70–80 GPa,
superhard carbides – silicon carbide α-SiC with a hexagonal structure and boron carbide B_4C with a trigonal structure,

superhard borides – tungsten boride WB with a tetragonal structure and hardness of 37.5 GPa and tetragonal titanium boride TiB_2 with hardness of 34.8 GPa,
superhard oxide – stishovite SiO_2,
superhard silicides – boron silicide B_4Si with a trigonal structure and hardness of 37–40 GPa,
superhard allotropic variety of carbon – diamond C with a regular structure and hardness of 100 GPa (polycrystalline diamond coatings with hardness of 63–99 GPa).

Stishovite SiO_2 is the only monooxide that meets the superhardness criterion. However, Badzian (1988) showed that it is possible to obtain mixtures of oxides that meet the given superhardness criterion. He described the synthesis of a material that is a mixture of boron oxides, e.g. B_2O, $B_{18}O$, $B_{22}O$, obtained from the B_2O_3 compound at a temperature of 1900–2300 K and a pressure greater than 5 GPa. The hardness of this mixture was at least 33 GPa.

The consequence of adopting the new superhardness criterion is that it is not met by, among others, single-phase binary materials that are compounds of nonmetals C and N with transition metals, i.e. monocarbides and mononitrides of transition metals from groups IVB to VIB (i.e. nitrides and carbides of: titanium, hafnium, zirconium, niobium, vanadium, tantalum, etc.), currently most often used for anti-wear coatings (Tables 1 and 2 elaborated on the base of Holleck 1984; Karczemska et al. 2008; Kupczyk 2015; Leger et al.; Novikov 1986; Vereshhaka et al. 1986; Wysiecki 1997). Their hardness in extreme cases does not reach 33 GPa, similarly to titanium carbide which has a hardness of 31.7 GPa according to Vereshhaka and Tretjakov (1986) or 32 GPa according to Hollek and Denker (1986), or tantalum nitride which has a hardness of 32.4 according to Olszyna (2001).

It should be noted that in the literature (Holleck 1984; Nishiyama 2012) the arrangement of superhard materials, in addition to the given pyramids, is also presented in the form of the so-called quadrilateral of the composition of superhard materials, described by Rediel, Kieffer and Benesowsky (Kupczyk 2015; Rediel 1994). It is based on four elements from group IIIA (boron), IVA (carbon and silicon), VA (nitrogen) of the periodic table of elements. Unlike the proposed hardness pyramid, it does not include oxygen, and therefore compounds in the form of oxides.

Table 1. Hardness of monocarbides of transient metals of IVB-VIB

Transition metal	Transition group	Monocarbide	Crystallographic structure	Hardness [GPa]
Ti	IV	TiC	fcc	32.0
Zr		TiC	cub-B1	31.7–32.0
Hf		ZrC	fcc	25.6
		ZrC	cub-B1	26.0
		HfC	fcc	22.0
		HfC	cub-B1	27.0
Nb	V	NbC	fcc	24.0
V		NbC	cub-B1	19.5–27.0
Ta		VC	fcc	29.5
		VC	cub-B1	23.4–27.6
		TaC	hcp	21,7
Mo	VI	MoC	hex	15.0
W		WC	hex	23.0–24.5

Table 2. Hardness of mononitrides of transient metals of IVB-VIB

Transition metal	Transition group	Mononitride	Crystallographic structure	Hardness [GPa]
Ti	IV	TiN	fcc	20.5–24.5
Zr		TiN	cub-B1	18.0–28.0
Hf		ZrN	fcc	20.0
		ZrN	cub-B1	13.0–20.0
		HfN	fcc	16.0–20.0
Nb	V	NbN	cub	14.61
VN		VN	cub	13.1
Ta		TaN	hcp	24.16–32.4
Cr	VI	CrN	cub	10.93
		CrN	fcc	18.0–21.0
		CrN	cub-B1	26.0–29.0

Carbon nitride C_3N_4 has been placed in the quadrangle of superhard materials. The hardness of this compound, not yet confirmed experimentally, is greater than the hardness of a regular diamond. Such assumptions result only from the calculations and analyses performed (Novikov 1986).

In the system presented in the quadrilateral of the composition of superhard materials, Badzian (1988), Burakowski (2004) and Olszyna (2001) found the possibility of formation of compounds lying on the line connecting C with BN by substituting pairs of C-C atoms with pairs of B-N atoms in the diamond structure under the pressure of 14 GPa and the temperature of 3000 K. Under these conditions, crystals of the $(BN)_xC_{1-x}$

type (for $0 < x < 1$) were obtained, with the composition of $(BN)_{0.60}C_{0.40}$, $(BN)_{0.26}C_{0.74}$ and $(BN)_{0.15}C_{0.85}$.

5 Conclusions

The proposed hardness system of superhard materials introduces order into the classification of superhard materials by arranging non-metals in the order of successive groups of the periodic table of elements from group III A to VI A. The adopted criterion of superhardness of materials may be a practical indication, among others, for manufacturers of coated tools, as it divides materials into hard ones, which can be directly deposited in the form of coatings on tools made of high-speed steels and other tool steels, and superhard ones, which are not suitable for effective use in the form of anti-wear coatings on the above-mentioned substrates without the use of transition layers due to their very limited ability to deform together with the substrate under the influence of significant unit pressures occurring in the cutting process. Transition layers suitably graded in terms of intermediate hardness "improve" the adhesion of the coating to the substrate. This was confirmed, among others, by the author's own studies [9]. In addition, it should be added that one of the most effective forms of using superhard materials is using them in the form of composite nano-crystalline materials. Their use should be expected to further increase the hardness of sinters made of superhard materials.

As indicated by numerous studies, including those conducted within the framework of my own research project (Kupczyk et al. 2011), the hardness of sintered materials can be increased by reducing the grain size. The use of nanograined ceramic material particles simultaneously leads to increased sinter ductility (Kupczyk et al. 2011; Olszyna 2001; Vaprek 1996; Wysiecki 1997).

The concept of superhard nanocrystalline composite sintered materials is based on the assumptions that the amorphization of the grain boundary structure significantly weakens the dislocation movements in this area, making it difficult to create plastic deformation, and that the number of structural defects created during the synthesis of the material decreases with the reduction of the grain size (Olszyna 2001; Trębiński 1992).

Examples of composites in which the reduction of grain size to nanometric size results in a significant increase in hardness are nanocrystalline nitrides of the MenN type with an amorphous grain boundary of the α-Si_3N_4 type. Hence, in the works (Olszyna 2001; Trębiński 1992) it is emphasized that in the studies on obtaining polycrystals of high hardness, the effort to reduce grain size should be combined with the modification of grain boundaries, especially in the case of nanocrystalline materials.

Acknowledgements. The author of the work would like to thank the Ministry of Science and Higher Education for the financial support of his own research project No. N N503 147734 (decision No.: 1477/B/T02/2011/34) and the project No. 0614/SBAD/1565 of the Institute of Mechanical Technology of Poznan University of Technology and for Dr. Andrzej Cellary for providing metrological equipment.

References

Badzian, A.R.: Superhard material comparable in hardness to diamond. Appl. Phys. Lett. **53**, 2495 (1988)

Burakowski, T.: Considerations on Synergism in Surface Engineering. Radom University of Technology Publishing House, Radom (2004)

Ciszewski, B., Przetakiewicz, W.: Modern materials in technology. Bellona Publishing House, Warsaw (1993)

Dobrzański, L.: Fundamentals of Materials Science and Metals Science, WNT, Gliwice–Warsaw (2002)

Holleck, H.: Binäre und tenäre Carbid-und Nitridsysteme der Übergangsmetalle, Berlin–Stuttgart (1984)

Karczemska A., et al.: Extraterrestrial, terrestrial and laboratory diamonds–Differences and similarities. Diamond Relat. Mater. **17**(7–10), 1179 (2008)

Khrushhov M., Babichev M.: Abrazivnoe iznashivanie, Moskva (1970)

Kieffer, R., Benesovsky, F.: Hartstoffe. Springer, New York (1963)

Kupczyk, M.J.: Surface engineering. Poznan University of Technology Publishing House, Poznan, Cutting tools (2015)

Kupczyk, M., Michalski, A., Żurek, J., Rosiński, M., Siwak, P.: Research on the technological and operational properties of cutting blades made of nano-carbides consolidated using pulsed plasma. Own Res. Project No. N N503 147734 (decision nr: 1477/B/T02/2011/34)

Leger, J.M., Haines, J., Szmidt, M., Petitet, J.P., Pereira, A.S.: Discovery of the hardest known oxide. Nature **383**, 401 (1996)

Nishiyama, N., et al.: Synthesis of nanocrystalline bulk SiO_2 stishovite with very high toughness. Scripta Mater. **67**(12), 955–958 (2012)

Novikov, L.W.: Sinteticheskie sverkhtviordye materialy. Naukova Dumka, Kijev (1986)

Olszyna, A.: Ceramika supertwarda. Publishing House of the Warsaw University of Technology, Warsaw (2001)

Rediel, R.: Novel ultrahard materials. Adv. Mater. **6**, 594 (1994)

Stishov, S.M., Popov, V.: Novaja plotnaja modifikacija kremnezjoma. Geokhimja **10**, 837 (1961)

Trębiński R.: Theoretical Basis for Designing Explosive Systems for Dynamic Synthesis of Superhard Materials, Bulletin WAT, Warsaw (1992)

Vaprek, S., Haussmann, M., Shizhi, L.: Electrochem. Soc. Proc. **96**, 619 (1996)

Vereshhaka, A.S., Tretjakov, I.P.: Rezhushhie instrumenty s iznosostojjkimi pokrytijami, Mashinostroenie, Moskva (1986)

Wysiecki, M.: Modern tool materials. WNT, Warsaw (1997)

Mechanical Properties of Biobased Composite with Treated Hemp Fibers in the Structural Reinforcement

Ivelina Ivanova(✉), Jules Assih, Cheikhna Diagana, and Isabelle Titeux-Peth

Université de Reims Champagne-Ardenne, Institut de Thermique, Mécanique, Matériaux (ITheMM, UR 7548), 51100 Reims, France
{ivelina.ivanova,jules.assih,cheikhna.diagana, isabelle.titeux}@univ-reims.fr

Abstract. The growing environmental concerns and the demand for sustainable materials have significantly increased interest in natural fibers as reinforcements in composite materials. Among these, hemp fiber stands out due to its excellent mechanical properties, biodegradability, and low environmental impact. This paper describes an experimental study of strengthening concrete structures by bonding hemp fiber fabrics with different adhesives (bio-based and synthetic epoxy). The main objective was to investigate the effect of strengthening with alkalization-treated and untreated hemp fiber fabrics. Treatment with NaOH (sodium hydroxide) destroys the link between the hydrogen and oxygen atoms in the hydroxyl groups of the fiber, replacing the hydrogen atom with a sodium ion. This reduces water absorption and increases structural integrity by removing structurally irrelevant components, like lignin and hemicellulose, from the hemp fiber. The study includes detailed tensile tests to assess the strength and the durability of hemp fiber composites and the bending tests of strengthened concrete specimens. Of course, test results are provided for both strengthened and unstrengthened concrete specimens, and their significance to member performance is also investigated. So, the measured load versus strain (with precision strain gauges) in composite hemp fiber fabric and concrete are plotted. Mainly, the focus is on measuring the effect of bonding and the usefulness of treating hemp fibers with a 5% NaOH solution to enhance their mechanical properties. The results show that the treated fibers increase the maximum bending load. The strength of strengthened concrete specimens by bonding hemp fiber fabrics with bio-based and synthetic epoxy after bending tests was compared and described.

Keywords: biobased composite · strengthening · reinforced concrete · mechanical behavior · hemp fiber fabrics · natural fibers

1 Introduction

The safety, durability, and longevity of reinforced concrete constructions is one of the most important tasks for engineers, of course. However, concrete inherently possesses excellent compressive strength but is limited by its relatively poor tensile strength, which

L. F. M. da Silva and R. D. Adams (Eds.): AB 2025, PEM, pp. 86–100, 2026.
https://doi.org/10.1007/978-3-032-19022-2_6

makes it susceptible to cracking and premature degradation under tension or dynamic loads. Strength of concrete structures by external bonding of composite material with natural fiber fabrics improves the mechanical properties, reduces the cracking, and significantly extends the lifespan of structures [1–4]. Moreover, with increased environmental considerations and resource constraints, the development of stronger, more durable concrete composites using sustainable materials is increasingly important. Enhancing the mechanical properties using natural fiber reinforcements, like treated hemp fibers, reduces the environmental footprint of construction materials. These improvements align with broader goals of sustainable development, resource efficiency, and reducing the carbon impact of the construction industry.

Natural fibers are an eco-friendly alternative to synthetic reinforcement in composite materials due to their biodegradability, renewability, and low density. Hemp fiber exhibits interesting mechanical properties, environmental compatibility, and cost efficiency, making it a strong candidate for structural reinforcement applications. However, for better compatibility with the polymer matrix, it is recommended to treat the fibers. Since untreated natural fibers tend to exhibit poor compatibility with polymer matrices due to their hydrophilic nature, high moisture uptake, and the presence of non-cellulosic components such as lignin, hemicellulose, and waxes [5, 6]. These characteristics reduce fiber-matrix adhesion and limit the mechanical performance and long-term durability of fiber-reinforced composites.

Chemical treatment allows modification to the fiber surface, reduces water affinity, and enhances interfacial bonding with polymeric or cementitious matrices. Among these, alkalization using sodium hydroxide (NaOH) is one of the most common and effective methods. This treatment removes amorphous components (like hemicellulose and lignin) and waxy substances from the fiber surface, increases surface roughness, and enhances mechanical interlocking. Studies have shown that the tensile strength of hemp fibers can increase from approximately 250–400 MPa (untreated) to 450–600 MPa after alkalization with 5% NaOH, accompanied by improved stiffness and reduced water absorption [7]. The choice of matrix also influences performance. Conventional epoxy resins are commonly used in research for natural FRPs due to their good mechanical properties and compatibility. Bio-based epoxies, made from renewable resources (plant oils, lignin, etc.), are an emerging sustainable alternative. The use of bio-based thermoset resins is one promising approach to reducing the large carbon footprint of petroleum-based resins. The use of bio-based epoxy resin of high-performance composites using jute fibers modified with graphene nanoplates (GNP) and graphene oxide (GO) demonstrated that physical and chemical treatments of jute fibers significantly improve their fiber volume fraction (Vf) and matrix adhesion, leading to enhanced mechanical properties of the resulting Jute/Bio-epoxy (J/BE) composites [8]. The use of bio-epoxy resins facilitates better adhesion of natural composite constituents [9]. In fact, both untreated and treated natural fiber composites can be made with either standard or bio-based epoxies, but treated fibers show clear advantages in tensile/flexural performance, interfacial bonding, and moisture resistance. [10]. Bio-based epoxy systems are improving rapidly and, when combined with fiber treatments, can approach the mechanical properties of traditional epoxy composites while being more sustainable [9, 11].

Composite laminates from natural fiber and epoxy, also called Natural Fiber Reinforced Polymer (NFRP), are a traditional external strengthening method for reinforced concrete structures. The natural fiber fabric is saturated with a standard epoxy resin and bonded to the concrete surface as an external laminate. Research in this area, particularly with jute and sisal fibers, shows that natural FRP wraps can significantly increase the load capacity and flexural strength of concrete members. The research of Tara Sen et al. studied the strengthened reinforced concrete beams in flexure with one layer of jute FRP (JFRP) and reported a 62.5% increase in ultimate bending strength compared to the unstrengthened beam [12]. By comparison, equivalent one-layer glass FRP gave 125% and carbon FRP 150% increase under the same conditions. Even with lower absolute strength, the jute FRP-strengthened beams showed favorable ductility: among the composites [12], indicating natural fibers allow more deflection and energy dissipation before failure. Fiber treatment further improves the effectiveness of NFRP laminates. Using pre-treated fibers in the laminate leads to a higher tensile capacity of the laminate. The better resin penetration, which provides a stronger connection and bond that can carry higher stresses [13]. The use of sisal fiber wraps found that externally bonded natural sisal FRP could raise the strength and ductility of concrete beams, though the percentage improvement depended on fiber quantity and anchorage [14]. To ensure adequate bonding to concrete, standard surface preparation (grinding, cleaning) is done as with any FRP, and bond-critical failures (debonding at sheet ends) can be mitigated by U-wraps or end anchors. Recent research specifically examined NFRP–concrete bond using various fibers and adhesives. It showed that bond strength increases with greater NFRP sheet thickness and width, and the use of a proper epoxy adhesive is critical [15]. Applying 2 layers of jute/abaca fabric (instead of 1) increased the FRP–concrete bond strength by roughly 118–350%, and 3 layers by 300% or more (depending on fiber type and resin) [15]. Externally bonded natural FRP laminates (especially with treated fibers) can meaningfully strengthen concrete – with reported flexural capacity gains on the order of 10–60% with one-layer strips [16] and higher with multiple layers – but they still lag behind high-strength synthetic FRPs. The trade-off is that they are cheaper, renewable, and have lower density (jute is 7 times lighter than steel fibers of equivalent length) [17]. Another work on accelerated aging indicates that while jute and sisal fibers do degrade in alkaline conditions, appropriate modifications can mitigate strength loss and preserve interface bonds [18].

This research presents an experimental study of strengthened structures by externally bonded composite fibers, as hemp and biobased adhesive, which is an effective solution for improving the strength of the structure. The main objective is to investigate their mechanical behaviour and, of course, the characteristics of cracking and failure modes. Ultimate load and the failure modes of the specimens are compared to the reference specimen. This experimental study contributes to the development of sustainable composites that have the potential to reduce the negative environmental impact of conventional materials while also offering improved mechanical properties.

2 Materials

2.1 Natural Hemp Fiber

Natural hemp fiber fabrics are attractive due to their low cost, low density, biodegradability, and availability. The hemp plant can be grown without artificial fertilizers. It covers the entire ground within 4 weeks. Growing it is easy, but the production is very labor-intensive, especially the separation of the fibers from the bast. During their cultivation, the hemp plant captures carbon dioxide; therefore, it is interesting for environmental protection. Hemp yarn is strong and has the highest water resistance, compared with other natural fibers.

The natural fibers have enhanced a matrix of biobased epoxy risen type to highlight their mechanical properties shown in Table 1.

The hemp fiber fabrics with meshes 3 × 3 mm were used, see Fig. 1. In Fig. 1a, the hemp fibre fabric is untreated, in contrast to Fig. 1b, where it was treated with Sodium hydroxide (NaOH, see Fig. 1b), with the main objective of improving their mechanical properties.

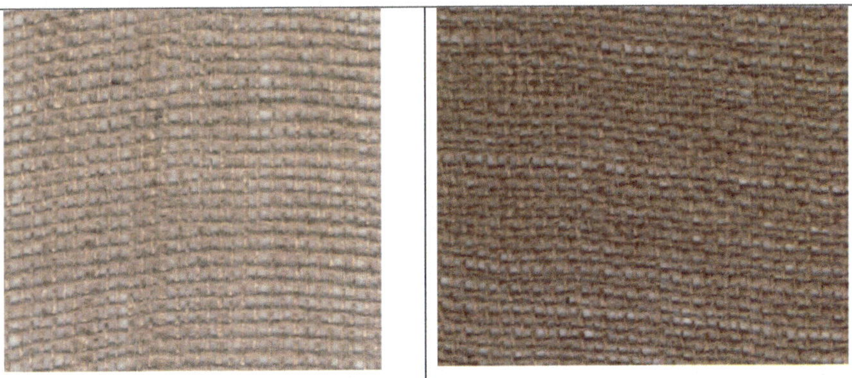

a) Hemp fabric of mesh of 3x3 mm

b) Treated Hemp fabric of mesh of 3x3mm

Fig. 1. The used fabric fibers

2.2 Fiber Treatment

Fiber surface treatments can significantly enhance the mechanical performance of natural fiber composites. The treatment removes amorphous components (like hemicellulose) and roughens or chemically activates the fiber surface, improving fiber-adhesives. As a result, treated fiber composites exhibit better stress transfer. The colour of the leftover NaOH solution after the treatment changed in colour. In Fig. 1b, a change in the color of the fibers is visible, as the sheets undergo a chemical reaction.

2.3 Concrete

Normal concrete was used for all prismatic concrete specimens. The cement: sand: gravel proportions in the concrete mix were 1:1.7:2.9 by weight, and the water/cement ratio was 0.5. The cement type II Portland was used. The maximum size of the aggregate was 16 mm. Four 160 mm × 320 mm concrete cylinders were cast and tested to determine the mechanical properties of the concrete at 28 days of age. The average compression strength was 35 MPa. The concrete had an average elastic modulus of 30 GPa. The biobased epoxy resin was used to make the composite. The mechanical properties of the adhesive are given as tensile stress of 50 MPa (see Table 1).

Table 1. Properties of the materials used and tested

Material	Young's modulus [GPa]	Strength [MPa]	Poisson ratio
Concrete	30	35	0.25
Biobased epoxy resin		50	
Treted Hemp 3x3 FCP*		362	

* FCP: Fiber Composite Plate.

3 Experimental Program

The experimental program examines the influence of the number of layers of composite of natural fibre fabrics using to external strengthening of the structures. Hemp fibre fabrics were used in this part of the study. However, using the hemp fibres, different samples were strengthened by composite sandwich fabrics were produced to assess the adhesion between the concrete and the panel, and the ultimate strength.

Finally, five types of composite layers were presented: 1, 2, 3, 4, and 5 layers of hemp fiber fabrics. Two series of five test tubes each are available, corresponding to the two types of adhesives, Sikadur 330 (traditionally used) and bio-based adhesive. The previous work allows us to rule out any inconsistency in the results that could be linked to a problem with the repeatability of our tests. Figure 2 shows the strengthening of concrete specimens according to the composite layer. The composites offer a combination of the advantages of constituent components to produce a material with determined properties. All dimensions are in centimeters.

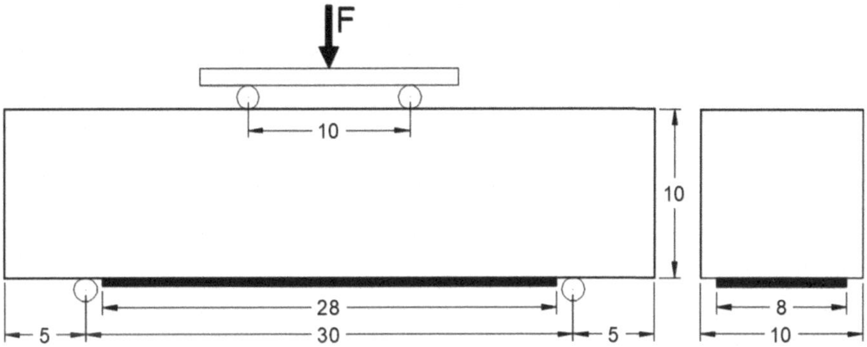

Fig. 2. Strengthening of concrete specimens according to the composite layer

The strength of the hemp fiber fabric composite is usually influenced by the composition of its constituent fibers. The mechanical properties change depending on the arrangement of the different layers of materials. Three specimens were tested for each configuration of strengthening at 28 days. All the produced specimens were examined in a 4-point bending test.

4 Results and Discussions

4.1 Influence of Different Adhesives

Figure 3 shows the mechanical behaviour of concrete structures reinforced with biosourced epoxy resin adhesive versus those reinforced with Sika structural adhesive. The elastic-plastic behaviour of the biobased adhesive is more obvious, with a threefold increase in deformation. This epoxy resin is 60% biosourced. The curing takes more time with biobased glue, even though it sets fairly fast. Of course, Sika adhesive has a brittle behaviour, while bio-sourced adhesive has an elastic-plastic behaviour. The choice of biobased glue appears to be an interesting alternative for reinforcing reinforced concrete structures.

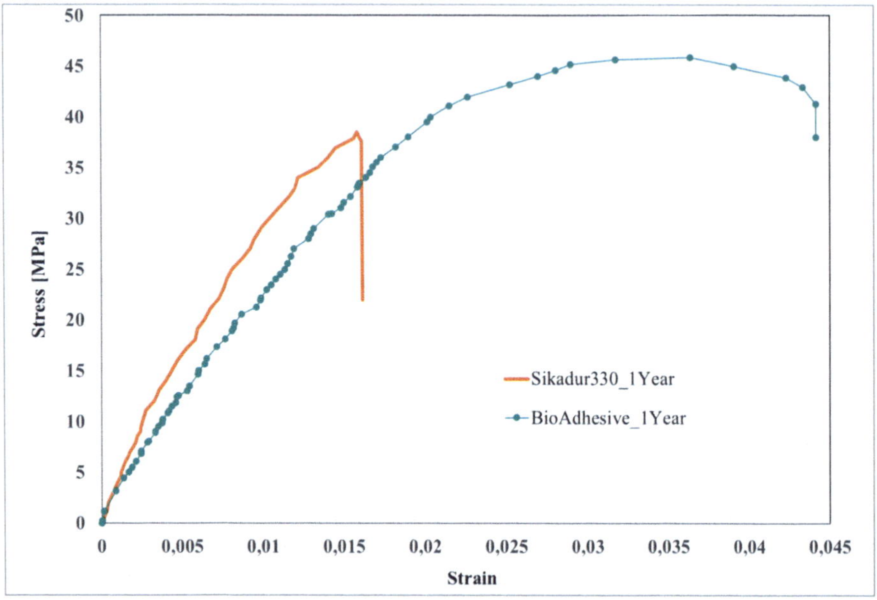

Fig. 3. Sika and biobased resin adhesive mechanical behavior in tensile test.

4.2 Aging Effect of Biobased Adhesives

A useful comparison is made in Fig. 4 of the behaviour of one year's aging under conditions of temperature 25 °C and average humidity 50%. The adhesive continues to harden over time, increasing its ultimate load-bearing capacity by more than a third.

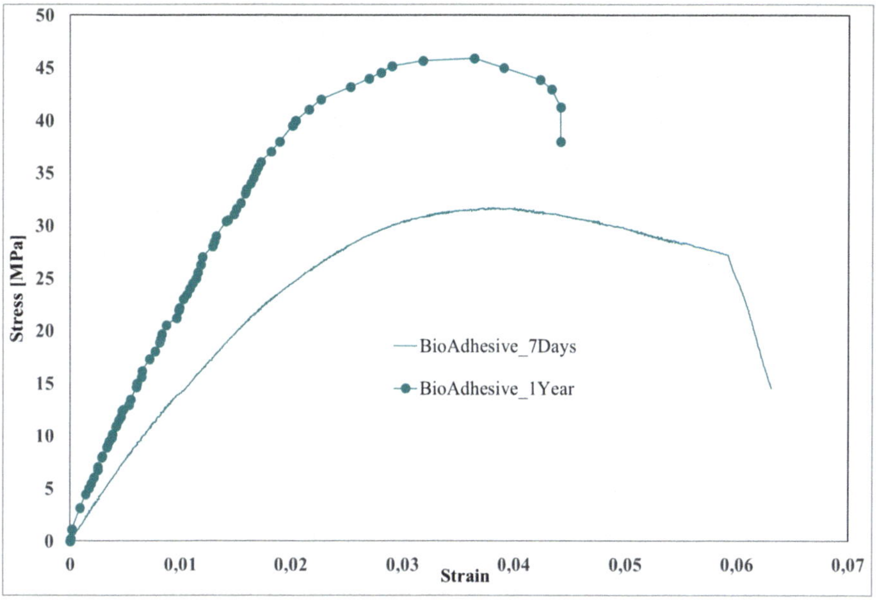

Fig. 4. Aging effect of Biobased adhesive

4.3 Influence of Fibre Fabrics on Ultimate Strength

4.3.1 Influence of the Number of Fiber Fabrics in the Sandwich Panel

The results showed an increase in the ultimate load from 57% to 163% according to the number of fiber fabrics. Mainly, two modes of failure were noticed: a bending crack in the concrete and the failure of composite fiber fabrics. The first results show an increasing variation in the ultimate strength of the samples. In consequence, it is not possible at present to define an optimum value for the ultimate load at failure. In so far as the predominant mode of collapse remains composite plate failure. These results, which always lead to the rupture of the libre composite sheet, led us to test sandwich sheets in order to avoid this type of collapse.

Table 2. Results of fiber fabrics number in the composite panel

Maximum ultimate load	Type of adhesive	Control specimen	Hemp FCP 1layer	Hemp FCP 2layers	Hemp FCP 3layers	Hemp FCP 4layers	Hemp FCP 5layers
F_u [kN]	Sika epoxy	12.09	19.02	20.70	19.00	23.95	31.85
F_u/F_0		0%	57%	71%	57%	98%	163%
F_u [kN]	Biobased epoxy resin	12.09	18.72	16.38	17.70	24.17	22.54
F_u/F_0		0%	55%	35%	46%	99.9%	86%

4.3.2 Influence of the Position of Fiber Fabrics of Hybrids Panels

Table 2 shows the strengthening by different layers of hemp fabric. The results show a collapse of the sample by a failure of the hemp fiber fabrics composite in the first two cases in the table. On the upside, in the cases of 3, 4 and 5 fabric layers, only a delamination of the composite plate at the end is observed.

4.4 Mechanical Behavior of Strengthened Sample Using Strain Gauges.

4.4.1 Extensometer Technique Based on Strain Gauges

Indeed, deformation versus applied load is used to study the local mechanical behaviour of the structure. Load is applied in equal increments, and at each increase, readings are recorded with those strains on composite fabrics, all data were at the middle of the concrete specimen. Deformation is measured by strain gauges of 120Ω and gauge factor of 2.09%. The strain gauges have a length of 5 mm and are used to measure strain on the composite plate glued on the concrete specimen.

4.4.2 Influence of Hemp Composite Thickness

The results show that the number of layers has little influence on the ultimate load in the case of strengthened concrete beams by epoxy resin adhesive. However, a phenomenon of creep between the different layers of hemp fibre fabrics is observed during this study. Figure 5 presents the mechanical behavior of the specimen, reinforced with a 5-layer fabric layer, which shows a perturbation at 10 kN followed by a steeper slope change until failure. The phenomenon of plaque debonding at the edges was observed mainly on the test piece with 5 layers. This can be clearly seen in the curves provided by the displacement captor. Nevertheless, the influence of the number of layers of fiber fabric on the behavior of strengthened concrete beams shows that the natural fiber fabrics have a linear elastic line up to the ultimate load. The effect of the NaOH treatment is visible in the force-displacement curves in Figs. 5 and 6. The stiffness was reduced by 25% and the ultimate load-bearing capacity by 30%. This result indicates that it is worthwhile to continue this project on large reinforced concrete beams.

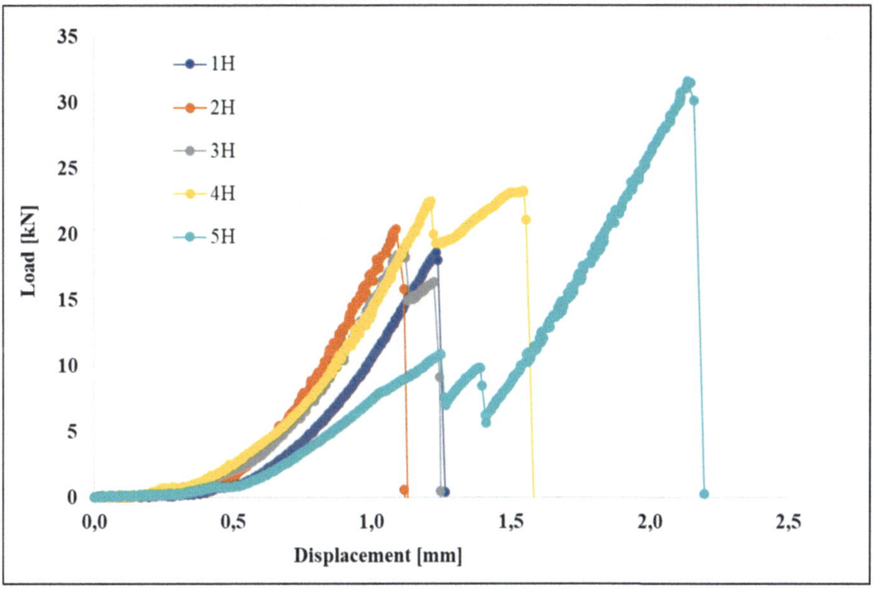

Fig. 5. Hemp thickness composite effect on the strengthened concrete beams

In fact, Fig. 6 shows the influence of NaOH-based hemp fibre treatment on the behaviour of a beam reinforced with different composite thicknesses. The results show an increase in ultimate load from a composite thickness of 4 layers upwards. For untreated fibres, on the other hand, this change is only observed from a thickness of 5 layers. This treatment effectively improves contact between the adhesive interface and the hemp fibre.

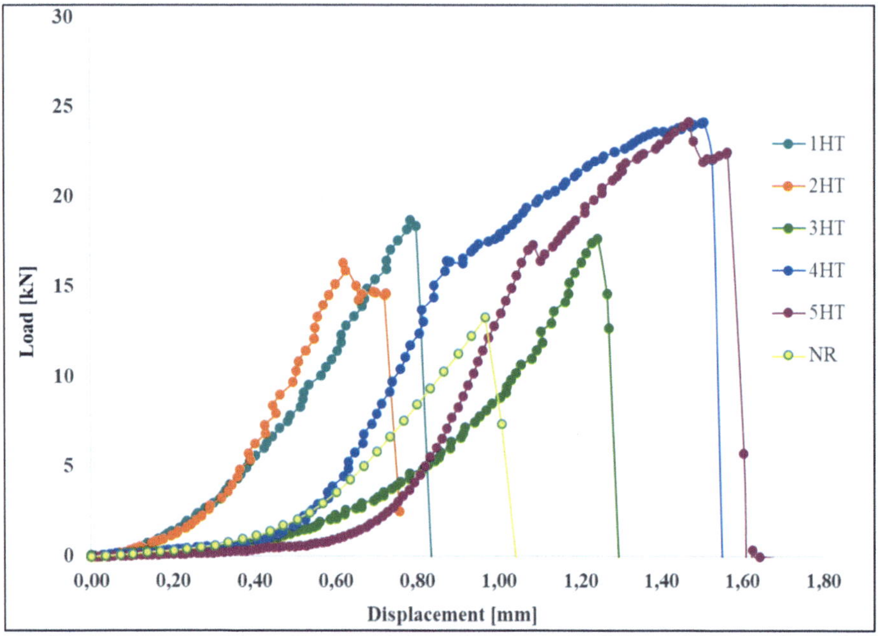

Fig. 6. Effect of hemp fibre fabrics treatment on the strengthening of concrete beams

Besides, Figs. 7 and 8 show the influence of composite thickness with data recording using strain gauges. Only, the results of hemp fiber fabrics treatment are shown. The results of the first three test tubes have been zoomed in on in Fig. 8 to see clearer. The strain of the reinforced beam, based on the number of treated hemp fibre layers. The two figures show the details of the strain, before and after fourth layer. Figure 8 shows a linear elastic section followed by sudden failure for beams 1HT, 2HT and 3HT. Both the 4HT and 5HT results show two distinct parts: elastic and non-linear. This partie provides ductility to the reinforced beam. The two figures show the details and confirm the change in mechanical behaviour from the fourth layer of hemp fibre. The strains are ten times greater from the fourth layer onwards.

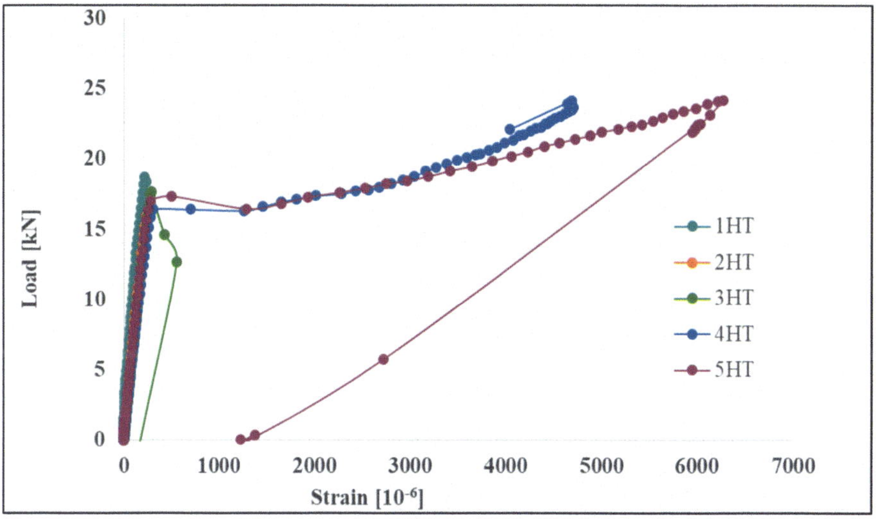

Fig. 7. Mechanical behavior of the strengthened concrete specimens by different layers of hemp treated fiber fabric.

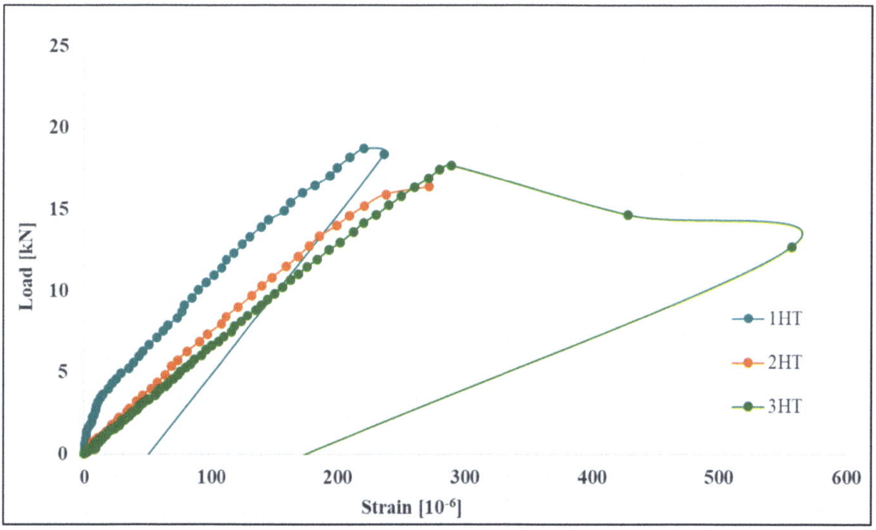

Fig. 8. Mechanical behaviour of the strengthened concrete specimens by different number of layers of hemp treated fiber fabric.

Table 3 shows the mode of collapse of strengthened concrete beams. Failure occurred after strengthening the concrete specimens. The results show that the composite plate breaks for several layers of untreated hemp fiber fabric of 4 and 3 for treated fibers. Beyond that, a disbound is observed at one end of the bonded plate.

Table 3. Mode of failure

	Specimen		Mode of failure
Control specimen			Bending crack in the concrete
1H			Failure of the composite plate
2H			Failure of the composite plate
3H			Failure of the composite plate
4H			Failure of the composite plate
5H			Non failure of the composite plate - Plate disbanding

5 Conclusion

This study reveals the potential of hemp fiber-reinforced composites with biobased epoxy resin as a sustainable and high-performance material for structural reinforcement.

The experimental results show a high potential for strengthening or repairing the reinforced concrete structures using adhesively bonded composite lamellas and increasing their service life. The results showed an increase in ultimate load from 35% to 99% in flexural strength compared to the control concrete beam without strengthening.

This experimental research approach provided a better understanding of the mechanical behavior and sliding techniques between fibers and glue. In our case, the treatment proved effective in changing the reinforced structure to make it more ductile. Of course, the ultimate load increases by around 10% and modifies the mode of failure and the cracking effect of different layers of the composite plate. The results of this project showed that the use of biobased adhesive in the strengthening of concrete beams increases its load bearing capacity significantly over time. The results show that the composite plate breaks for several layers of untreated hemp fiber fabric of 4 and 3 for treated fibers. Beyond that, a disbound is observed at one end of the bonded plate.

Acknowledgments. The authors gratefully acknowledge the financial support from the University of Reims Champagne Ardenne and ITHEMM. Thanks also to the support provided by the Bulgarian Foundation of Scientific Research, and UCTM supported this work under contract КП-06-H57/13-926, 2021-Mechanical study of strengthened reinforced concrete structures by natural fiber fabrics under dynamic load.

References

1. Ivanova, I., Assih, J., Dontchev, D.: Investigation of the mechanical behavior of natural vegetable fibers used in composite materials for structural strengthening. Key Eng. Mater. **888**, 15–21 (2021). https://doi.org/10.4028/www.scientific.net/KEM.888.15
2. Assih, J., Ivanova, I.: Experimental and analytical study of bio-based epoxy composite materials for strengthening reinforced concrete structures. Sci. Eng. Compos. Mater. **28**(1), 335–342 (2021). Open Access published by De Gruyter. https://doi.org/10.1515/secm-2021-0024
3. Bou Abdallah, G., Ivanova, I., Assih, J., Diagana, C., Dontchev, D.: Composite materials based on natural fibers applied for structural reinforcement. IOP Conf. Ser.: Earth Environ. Sci. **960**(1), 012007 (2022)
4. BouAbdallah, G., Diagana, C., Assih, J., Titeux-Peth, I., Ivanova, I., Dontchev, D.: Strengthening of structures by composite materials based on jute and flax fibers. Acad. J. Civ. Eng. **40**(1), 211–214 (2022). https://doi.org/10.26168/ajce.40.1.52, Special Issue-RUGC 2022
5. Samanth, M., Subrahmanya, B.K.: Conventional and unconventional chemical treatment methods of natural fibres for sustainable biocomposites. Sustain. Chem. Clim. Action **3**, 100034 (2023). https://doi.org/10.1016/j.scca.2023.100034
6. Ali, A., et al.: Hydrophobic treatment of natural fibers and their composites—A review. J. Ind. Text. **47**(8), 2153–2183 (2016). https://doi.org/10.1177/1528083716654468
7. Shahzad, A.: Hemp fiber and its composites–A review. J. Compos. Mater. **46**(8), 973–986 (2012). https://doi.org/10.1177/0021998311413623
8. Islam, M.H., Afroj, S., Karim, N.: Toward sustainable composites: graphene-modified jute fiber composites with bio-based epoxy resin. Glob. Challenges (Hoboken, NJ) **7**(9), 2300111 (2023). https://doi.org/10.1002/gch2.202300111
9. Capretti, M., Giammaria, V., Santulli, C., Boria, S., Del Bianco, G.: Use of bio-epoxies and their effect on the performance of polymer composites: a critical review. Polymers **15**(24), 4733 (2023). https://doi.org/10.3390/polym15244733
10. Luhar, S., et al.: Sustainable and renewable bio-based natural fibres and its application for 3D printed concrete: a review. Sustainability **12**(24), 10485 (2020). https://doi.org/10.3390/su122410485
11. Ghorbel, E., Limaiem, M., Wardeh, G.: Mechanical performance of bio-based FRP-Confined recycled aggregate concrete under uniaxial compression. Materials **14**(7), 1778 (2021). https://doi.org/10.3390/ma14071778
12. Sen, T., Jagannatha Reddy, H.N.: Strengthening of RC beams in flexure using natural jute fibre textile reinforced composite system and its comparative study with CFRP and GFRP strengthening systems. Int. J. Sustain. Built Environ. **2**, 41–55 (2013). https://doi.org/10.1016/j.ijsbe.2013.11.001
13. Sen, T., Jagannatha Reddy, H.N.: Pretreatment of woven jute FRP composite and its use in strengthening of reinforced concrete beams in flexure. Adv. Mater. Sci. Eng. **128158**, 15 (2013). https://doi.org/10.1155/2013/128158
14. Song, H., Liu, J., He, K., Ahmad, W.: A comprehensive overview of jute fiber reinforced cementitious composites. Case Stud. Constr. Mater. **15**, e00724 (2021). ISSN 2214-5095. https://doi.org/10.1016/j.cscm.2021.e00724

15. Saidi, T., Hasan, M., Amalia, Z., Salsabila, S.: The analysis of the bond strength between natural fiber reinforced polymer (NFRP) sheets and concrete. Results Eng. **18**, 101124 (2023). ISSN 2590-1230. https://doi.org/10.1016/j.rineng.2023.101124
16. Salih, Y.A., Sabeeh, N.N., Yass, M.F., Ahmed, A.S., Khudhurr, E.S.: Concrete beams strengthened with jute fibers. Civ. Eng. J. **5**(4), 767–776 (2019). https://doi.org/10.28991/cej-2019-03091286
17. Ahmad, J., et al.: Performance of concrete reinforced with jute fibers (natural fibers): a review. J. Eng. Fibers Fabr. **17** (2022). https://doi.org/10.1177/15589250221121871
18. Jamshaid, H., Ali, H., Mishra, R.K., Nazari, S., Chandan, V.: Durability and accelerated ageing of natural fibers in concrete as a sustainable construction material. Materials (Basel, Switzerland) **16**(21), 6905 (2023). https://doi.org/10.3390/ma16216905

Manufacturing

Joining Adhesive Bonds - Part 1: Characterization of Non-newtonian Materials - A Basis for Squeeze Flow Simulations

Franziska Hagemann[1(✉)], Torsten Remmler[2], and Manuel Schiel[1]

[1] Henkel AG & Co. KGaA, Henkelstraße 67, 40589 Düsseldorf, Germany
{franziska.hagemann,manuel.schiel}@henkel.com
[2] NETZSCH-Gerätebau GmbH, Wittelsbacherstraße 42, 9500 Selb, Germany
torsten.remmler@netzsch.com

Abstract. Adhesives and thermal interface materials play a critical role in industrial production, particularly in the shift from internal combustion engines to battery electric vehicles. These materials are essential for battery module assembly, serving functions such as sealing housings and facilitating thermal management. Thin and fragile components must be joined without damage. A key factor in this process is understanding the squeeze force required to achieve the final gap height during assembly. Simulations can predict squeeze flow behavior under various conditions, offering a valuable supplement to traditional squeeze flow experiments. Simulations still rely on experimental data from which the material model is derived. Since most adhesives exhibit non-Newtonian flow behavior, accurate prediction of the joining force requires detailed rheological characterization. This study presents a systematic approach to determine shear viscosity functions through rheological experiments. The focus is on optimizing rheological testing methodologies to derive reliable viscosity functions tailored to adhesive joining processes, thereby enhancing the accuracy and efficiency of simulation-based assessments.

Keywords: Non-Newtonian Adhesive Materials · Rheology · Material Model · Squeeze Flow

1 Introduction

Adhesives play a critical role in a wide range of industrial and consumer applications, from aerospace and automotive assembly to electronics and medical devices (Wei et al. (2024)). Especially for the transition of internal combustion engines to battery electric vehicles, adhesive materials provide concepts for bonding and an efficient thermal management.

During battery pack assembly, the liquid adhesive is dispensed onto one part and subsequently, the adherents are joined. Several phenomenons are of interest during this joining process: The determination of the forces needed to join the

L. F. M. da Silva and R. D. Adams (Eds.): AB 2025, PEM, pp. 103–117, 2026.
https://doi.org/10.1007/978-3-032-19022-2_7

two parts until the final gap height is reached, the evaluation of how the adhesive spreads and how to avoid air enclosures, as well as the reduction of material waste. The joining process, described as a squeeze flow, is defined by the squeeze velocity, the application pattern and most importantly by the adhesive's flow behavior (Aparecida Silva et al. (2022)). The latter is heavily impacted by the adhesive's rheological properties, resulting from the adhesive's composition.

In literature many publications study how to calculate squeeze flow analytically and numerically (Engmann et al. (2005); Aparecida Silva et al. 2022)). To this end, a mathematical material model, such as the Herschel-Bulkley (Herschel and Bulkley (1926)), Carreau (Carreau (1972)), or Power-Law model (Waele (1926)), is necessary to describe the flow behavior of the squeezed adhesive. The model is derived from rheology measurements, in particular viscosity measurements, designed in a way that the shear rate development is similar to the one in the squeeze flow gap. The rheology of adhesives is complex, due to numerous ingredients such as different polymers or filler particles. The latter are added to materials to reduce cost, modify mechanical properties, increase viscosity, improve electrical, or thermal conductivity (Sanghvi et al. (2022)). Filled materials were extensively studied in literature (Genovese (2012); Kovačević et al. (1994); Lemke et al. (1999); Rueda et al. (2017); Barnes (2003)). It was found that the filler content heavily impacts the material's rheology and can also cause thixotropy and wall slip effects, which make the viscosity analysis for the determination of material models challenging. A method that accounts for the correction of wall slip in a parallel plate rheometer set-up was presented by Yoshimura and Prud'homme (1988). Mooney (1931) developed a methodology for wall slip correction in capillary rheometers. Some materials exhibit a yield point. They are able to flow only if they are exposed to shear stress above a critical stress. Below this value they behave like solids. Coussot (2014) reviewed different experimental methods to determine non-thixotropic yield points. Further yield point analysis was performed by Ong et al. (2019) and Ovarlez et al. (2013).

Recent publications on squeeze flow are focusing on the visualization of the adhesive distribution (Kaufmann et al. (2022; 2023; 2024)) and on numerical methods to describe squeeze flow (Müller et al. (2023); Huf et al. (2024)). Automotive companies extensively investigated the rheology of adhesives to understand effects during dispensing, using capillary rheometry (Li et al. (1990); Hicks et al. (2015); Kamar and Hill (2024)). The aim of this study is to define the most suitable rheological test settings for squeeze flow simulations, as the rheological data is the foundation on which the simulations are based. This study explores the rheological behavior of adhesives using parallel plate flow and capillary rheometer measurements. It is the aim to establish test conditions as close as possible to the ones occurring in the narrowing gap. Through this analysis, we aim to highlight the importance of the appropriate rheological evaluation for the presented use case. It should help to understand the complex behavior in squeeze flow applications and could be extended to predict adherent deformation or adhesive distribution in the fluid state.

2 Material and Methods

2.1 Materials

In this work two different materials, used in batteries for battery electric vehicles are investigated. Both are two component materials.

- Two-component adhesive (adhesive and booster, mixing ratio 40:1)
- Two-component thermal gap filler (part A and part B, mixing ratio 1:1)

Both materials are highly filled. The fillers have a broad particle size distribution, with $d_{90} = 100\ \mu m$. Particles are not perfectly spherical and particle agglomerates are possible. The filler particles in the thermal gap filler enable heat transfer. In the polyurethane based adhesive the filler material is incorporated to increase the viscosity. For the considered materials, the viscosity's impact superimposes the impact of the contact angle, and is therefore the parameter to describe the flow characteristics of a material.

2.2 Methods

Assuming a constant squeeze velocity, the adhesive's flow velocity and with that the shear rate increase with narrowing gap. This is why also rheology measurements are to be run from low to high shear rates. For squeeze flow applications, rotational and capillary rheometers cover the necessary shear rate range (Da Silva et al. 2011)). All experiments were conducted at 23 °C and each experiment was repeated three times. The error bars were smaller than the symbols in the graph.

2.2.1 Viscosity and Stress Determination

Rheometers

The viscosity curve for shear rates from $2{\cdot}10^{-4}$ to approximately $10^{1/s}$ was determined with a shear rate-controlled parallel plate rheometer (ARES-G2 from TA Instruments). A plate diameter of 25 mm was used for the adhesive (+ booster) to reduce the axial forces needed for reaching the final gap height. For the thermal gap filler less compression force was needed and therefore 40 mm plates were used. The plate surface was adapted to the investigated materials, avoiding wall slip and wall effects. The shear rate was increased stepwise and the viscosity was determined after steady state was reached for the torque versus time function. At a certain shear rate the materials showed edge effects due to centrifugal forces or elastic normal stresses under shear flow, known as edge failure. Measurements for shear rate above this limits were performed with a capillary rheometer (RH2200 from NETZSCH-Gerätebau GmbH).

Thixotopy Effects

To prevent thixotropic memory effects of the adhesive materials each sample was subject to a rejuvenation protocol before starting the respective rheometry measurement in the parallel plate rheometer. A constant shear rate of $0.1^{1/s}$ was applied to the sample in the rotational plate rheometer for 60 s which was sufficient to reach steady-state shear flow. These settings are valid for both investigated materials.

Curing
The two-component adhesives should be measured in their mixed state to capture the most realistic viscosity function, whenever possible. If the time required to measure the viscosity curve exceeds the onset time of the curing reaction, the viscosity of the material may be affected during the measurement. Or the material cures completely, making viscosity measurements impossible. Adhesive and booster were mixed in a ratio of 40:1 in a centrifugal mixer. Due to the energy input into the system, the temperature of the mixture increases, entailing an undefined temperature field in the sample during the measurements. Furthermore, the mixture changes its characteristics over time due to curing. For the mixture of adhesive and booster the skin formation time was 25 min. To prevent the equipment to stick together shear rates below $0.01^{1/s}$ were not investigated as the equilibration time is too long. Also, measurements with the capillary rheometer for high shear rates were not possible due to the measurement and equipment cleaning time. This is why the viscosity curve for the mixture was only measured in the shear rate range from 10^{-2} to $100^{1/s}$. Further viscosity measurements were performed only with the adhesive (without booster) to avoid changes in the adhesive's rheology over time. To eliminate curing reactions due to air humidity, rheology experiments in the parallel plate rheometer were performed under nitrogen atmosphere.

Wall Slip
Wall slip can occur in rheometers resulting in a reduced measured viscosity. To determine if wall slip is present, the measurement in the parallel plate rheometer was repeated with different gap heights. For the adhesive smooth plates with a diameter of 25 mm were used and the measurements were performed with gap heights of 0.8, 1.0, and 1.1 mm. As no wall slip was detected no further actions were required. For the thermal gap filler parallel plate measurements smooth and sand blasted plates of 40 mm diameter were used. The surface roughness of the sand blasted plates was Sa = 6.4 μm. The following gap heights were investigated with smooth plates: 1.0, 1.5, and 1.7 mm. As wall slip was detected, experiments were repeated with sand blasted plates. Gap heights of 0.5, 1.0, 1.5, and 1.7 mm were tested. In the capillary rheometer the following die configurations were used: $d_1 = 1$ mm, $L_1 = 16$ mm and $d_2 = 1.5$ mm, $L_2 = 24$ mm. A Mooney correction was not necessary, as wall slip did not occur in the capillary rheometer (Mooney (1931)).

2.2.2 Checking for Yield Stress

The same plates, gap heights, and rejuvenation protocols as determined in Sect. 2.2.1 were used for the yield stress pre-tests. To determine whether a material has a yield point several conditions have to be fulfilled:

1. In rotational steady-state shear flow measurements the shear stress runs into a plateau for shear rates towards zero: The shear stress course is determined as described in Sect. 2.2.1. If the plateau condition for small shear rates was fulfilled, oscillatory measurements were started with the shear rate-controlled rotational plate rheometer.
2. In oscillatory frequency measurements within the linear visco-elastic region (LVR) G′ must be smaller than G″. G′ and G″ must both run into a plateau with decreasing the frequency: In the LVR, the strain does not impact the micro-structure of the material and deformation is reversible. The oscillation strain was increased from 10^{-3} to $10^{2}\%$ for the two frequencies: 0.1 and 10 rad/s, representing the upper and lower frequency limit. The end of the LVR was fixed to the strain where the derivations of G′ and G″ according to the shear strain amplitude were ≤ 0.1. This strain value was then used for the oscillatory frequency sweep.
3. In oscillatory strain measurements, the phase angle drops below 45° for angular frequencies towards zero: For phase angles <45° the storage modulus G′ is larger than the loss modulus G″. This means, that the material's network structure can only be set in motion above a critical shear stress - the yield stress.

When all three conditions were fulfilled, the yield stress was deviated from the plateau in the shear stress over shear rate diagram. This value was then taken as an initial value for the material model parameter fitting.

2.3 Material Models

In order to simulate the behavior of non-Newtonian fluids, a suitable material model is needed. The adhesive, which showed a yield point, was described by the Herschel-Bulkley-Papanastasiou model (Papanastasiou (1987)):

$$\eta = \frac{\tau_{\dot{\gamma}}}{\dot{\gamma}} \cdot (1 - \exp(-m \cdot \dot{\gamma})) + K \cdot \left(\frac{\dot{\gamma}}{\dot{\gamma}_{\text{ref}}}\right)^{n-1} \tag{1}$$

The gap filler was assumed to behave like a zero-shear plateau and was modeled by the Carreau model (Carreau (1972)):

$$\eta = \eta_{\text{inf}} + (\eta_0 - \eta_{\text{inf}}) \cdot \left(1 + (\lambda \cdot \dot{\gamma})^2\right)^{\frac{n-1}{2}} \tag{2}$$

3 Results and Discussion

3.1 Rheological Investigation

3.1.1 Adhesive

Due to curing, the mixture of adhesive and booster was not investigated in the full shear rate range of interest. The adhesive alone was used to determine the shear viscosity function needed for the material model. Figure 1a shows that the shear viscosity of the mixture is very similar to the one of the adhesive in the evaluable range. The rheological properties of the adhesive do not exhibit significant changes that alter the viscosity curve within the examined shear rate range when adding the booster. From these results it is assumed that also lower and higher shear rates are dominated by the characteristics of the adhesive and all subsequent investigations are performed with it.

In Fig. 1b oscillatory and rotational viscosity measurements are compared. The results show clearly that the Cox-Merz rule is not applicable for the highly filled adhesive, which is in accordance with literature (Doraiswamy et al. (1991); Gleissle and Hochstein (2003)). This is why rotational measurements were used in the following.

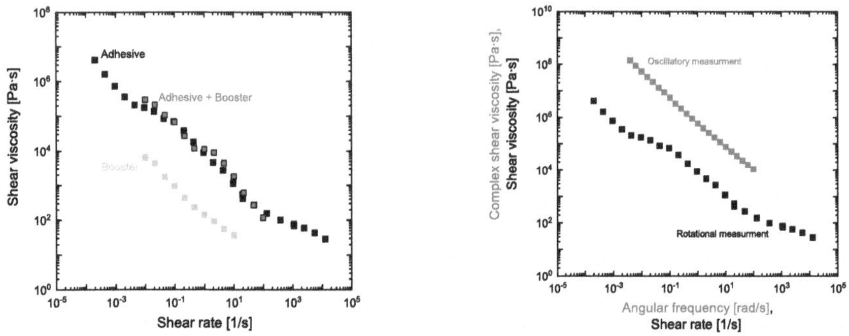

(a) Viscosity functions of the mixture and the components. Mixture and adhesive behave very similar in the considered shear rate range.

(b) Comparison of rotational and oscillatory adhesive measurement: Cox-Merz rule is not applicable to the investigated material.

Fig. 1. Shear rate dependent viscosity of the filled adhesive. Parallel plate rheometer: 10^{-4} 1/s $< \dot{\gamma} < 1^{1/s}$ and capillary rheometer $\dot{\gamma} > 1^{1/s}$.

Next, it was investigated, if the adhesive exhibit wall slip. For squeeze flow simulations a viscosity function with defined boundary conditions at the wall is needed and wall slip has to be eliminated. Figure 2a shows that in parallel plate experiments the gap height has no significant impact on the shear viscosity curve, signifying that wall effects can be neglected and a no-slip condition is present.

The shear viscosity shows a linear decreasing trend with increasing shear rates. However, there is a bump visible at a shear rate of $0.1^{1/s}$ in the viscosity curve. It manifests as a plateau in the corresponding stress curve shown in Fig. 2b, indicating structural changes at the molecular level. Another plateau in the stress curve is located at a shear rate of $0.03^{1/s}$. Plateau in the tress curve and a linear decreasing shear viscosity curve are both conditions for a yield point. In the following this fact is verified, checking the LVR, the course of the phase angle, and the frequency sweep.

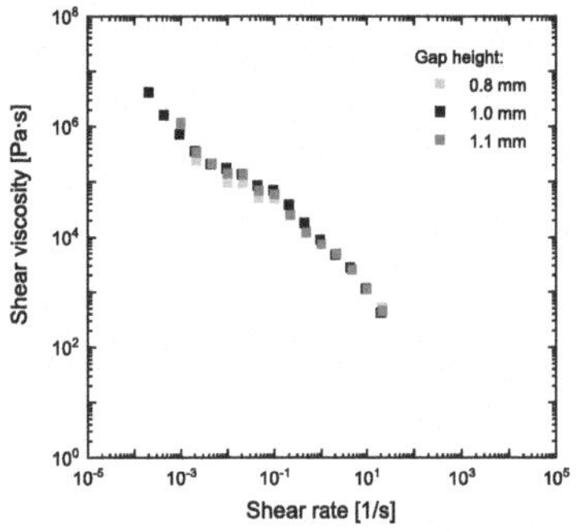

(a) Investigation of wall effects: Different gap heights show the same viscosity course, which means that wall effects are negligible.

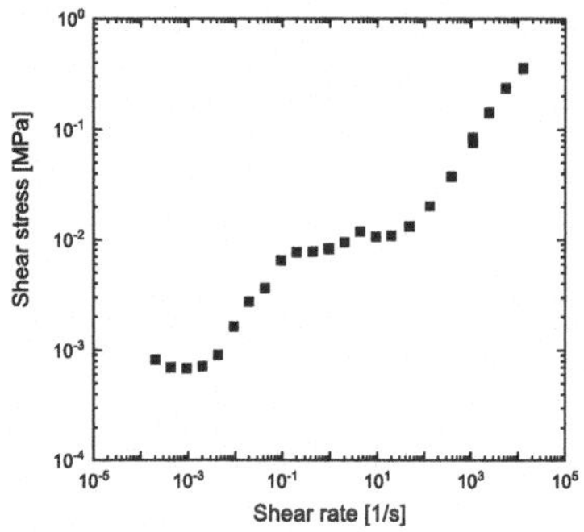

(b) Stress curve over the shear rate. The plateau for small shear rates indicates that the material might have a yield point.

Fig. 2. Viscosity and stress curve over the shear rate for the adhesive without booster.

The LVR of the material, cf. Figs. 3a and 3b, is located in the plateau region of the shear modulus at lower strain amplitudes. For frequencies from 0.1 to 10 rad/s the adhesive shows linear viscoelastic behavior for a shear strain amplitude of $\leq 0.05\%$. The frequency sweep, which was performed in the LVR at a shear strain 0.05%, exhibits a plateau for low angular frequencies, cf. Fig. 3c. The phase angles equally run into a plateau below 45° for decreasing shear strain, cf. Fig. 3d. A phase angle <45° means that the elastic shear modulus G' is greater than the viscous shear modulus G''. If this condition applies for circular frequencies approaching zero, then there is a network structure that can only be converted into flow when a shear stress limit is exceeded. A the yield stress of approximately 700 Pa can be derived from the lower plateau shown in Fig. 2b.

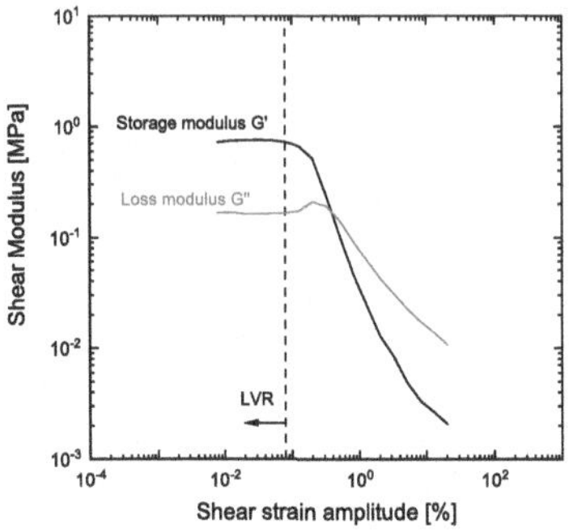

(a) Strain sweep of the adhesive for a frequency of 0.1 rad/s.

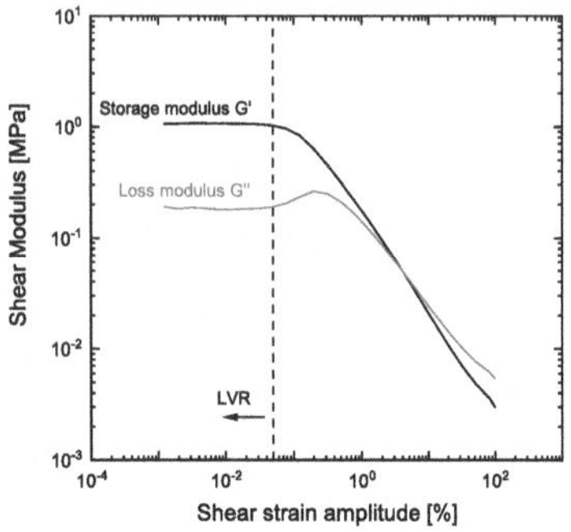

(b) Strain sweep of adhesive for a frequency of 10 rad/s.

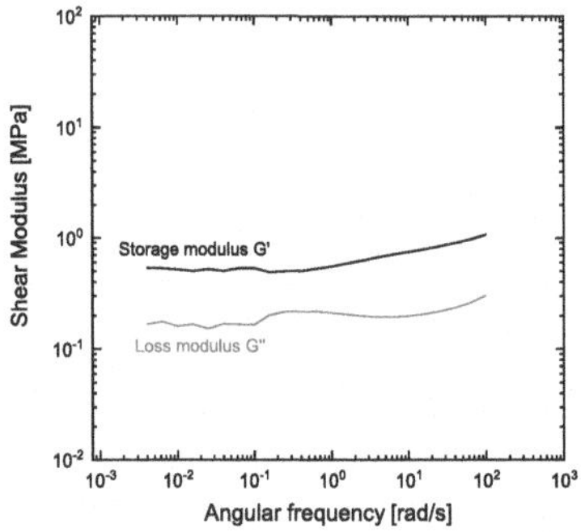

(c) Angular frequency sweep at an oscillation strain of 0.05% .

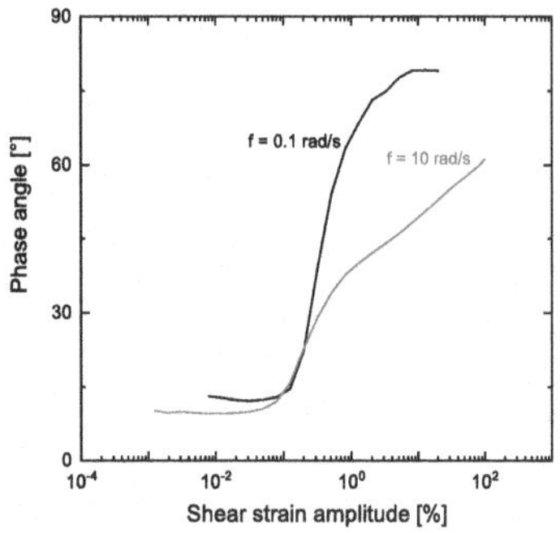

(d) Phase angle for evolution for the angular frequencies 0.1 rad/s and 10 rad/s.

Fig. 3. Investigation of the linear viscoelastic region, the angular frequency sweep and the phase angle evolution of the adhesive: All of these measurements indicate that the material has a yield point.

3.1.2 Thermal Gap Filler

This section discusses the rheology of the highly filled thermal gap filler. First, the impact of the gap height on the viscosity was investigated using 40 mm diameter smooth plates, cf. Fig. 4a. The results show, that the viscosity is gap height dependent which is caused by wall effects. Barnes (2003) summarized the impact of the filler content on the adhesive's rheology, pointing out that interacting filler particles can cause depletion or flocculation leading to wall slip effects. Smooth surfaces, horizontally aligned systems, materials with large particles, and low shear rates are most vulnerable for wall depletion effects and with that for wall slip (Barnes (1995)). For precise results of squeeze flow simulations it is important to provide material models from defined shear viscosity data. When wall slip is present, the measured viscosity curve is reduced due to lower reaction forces at the wall. This would underestimate the squeeze force in squeeze flow simulations and is the reason why viscosity data should be measured without wall slip. Figure 4b depicts the viscosity curve for 40 mm diameter sand blasted plates, showing no impact of gap height and wall effects. For visualization, the wall adhesion on the smooth and the sand blasted plates is compared in Fig. 5, showing slip over a time frame of 5 min on the smooth plate. On the sand blasted plate, the material remains unchanged. This behavior is in accordance with the findings of Meeten (2004).

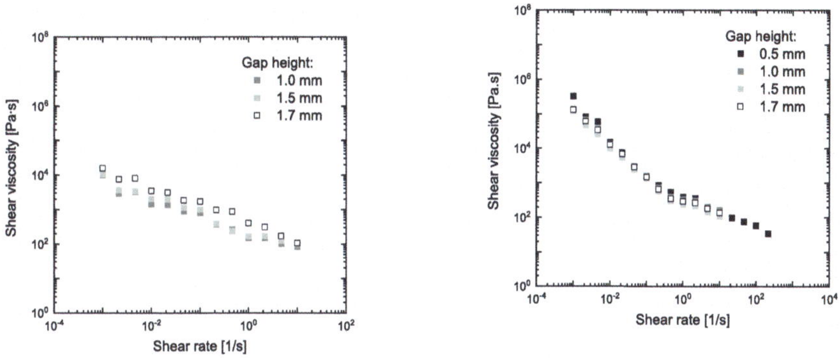

(a) Smooth plate with 40 mm diameter.

(b) Sand blasted plate with 40 mm diameter.

Fig. 4. Viscosity curve for the thermal gap filler: The impact of wall slip is clearly visible for a smooth plate set-up.

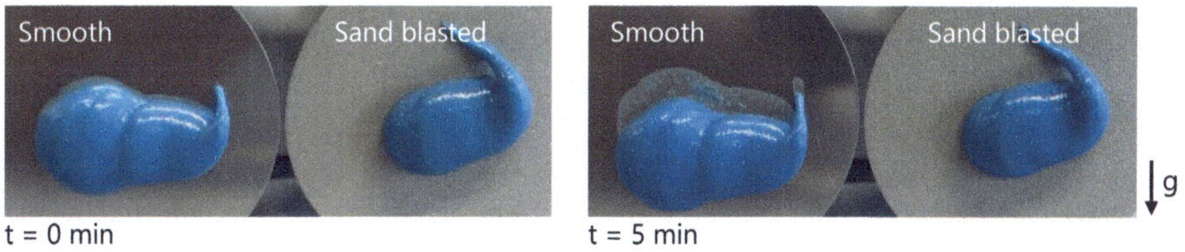

Fig. 5. Impact of the surface on the adhesion of the thermal gap filler on the wall. They were positioned vertically and the gravitation acts in the direction of the arrow: On the smooth plate, the material slides downwards, whereas it is stable on the sand blasted surface.

It was also investigated whether wall slip impacted the capillary rheometer measurement, cf. Fig. 6. The two shear stress curves measured with different die geometries of the same diameter to length aspect ratio form a straight line, indicating that no-slip conditions are present in the die capillary, cf. Mooney

(1931). The reason why there is a no-slip condition and also higher shear rates in the capillary and wall slip in the parallel plate rheometer could be due to a rougher surface in the dies of the capillary rheometer. Furthermore, the capillary rheometer is vertically aligned. According to Barnes (2003) these systems are less vulnerable for wall slip than horizontally aligned systems, such as parallel plate set-ups.

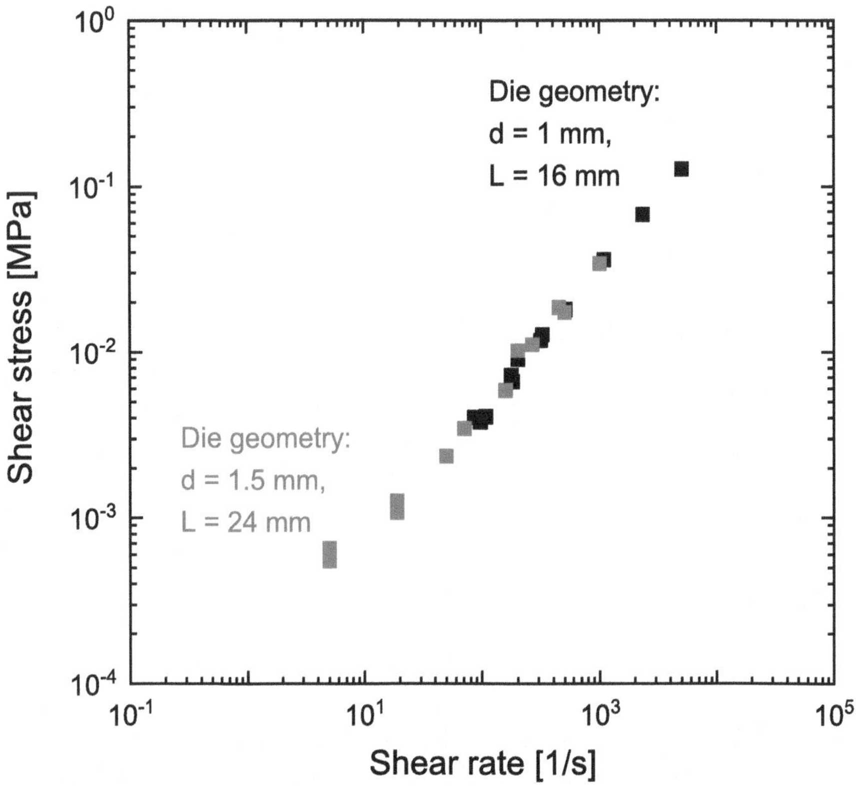

Fig. 6. Investigation of wall slip in the capillary rheometer: The two stress curves, measured with dies of the same diameter/length aspect ratio form a straight line which means that there are no-slip condition at the die wall.

Shear viscosity and shear stress for the thermal gap filler are presented in Figs. 7a and 7b, respectively. In both graphs, the mixture is compared to the two components A and B. It can be concluded that the viscosity characteristics of component A are prevailing in the mixture as both, mixture and component A have a similar viscosity progression. The shear viscosity function of the mixture is independent of the mixing ratio, which was in this case 1:1. The stress curve for component B has a plateau region at small shear rates indicating yield stress behavior. Component A and the mixture, however, exhibit an unconventional behavior: With increasing shear rates the shear stresses for both materials first run through a minimum before they increase linearly. As the data was measured in steady state, this behavior might be caused by the high ratio of filler particles and has to be further investigated in the future.

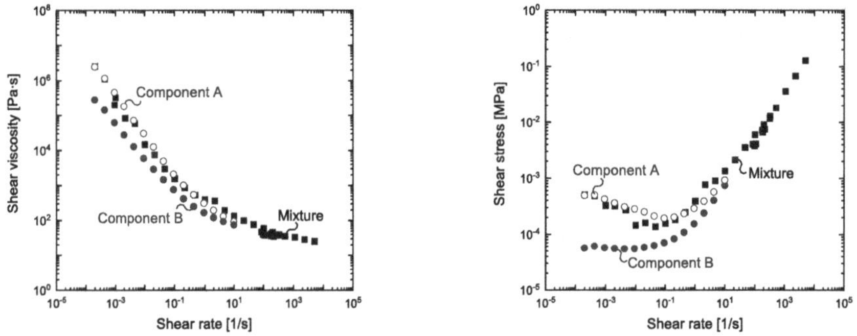

(a) Viscosity curve.

(b) Stress curve.

Fig. 7. Viscosity and stress curve comparison of the mixture and the two components of the thermal gap filler. The mixture and the single component A behave very similar in the considered shear rate range. Parallel plate rheometer: 10^{-4} 1/s $< \dot{\gamma} < 100^{1/s}$ and capillary rheometer $\dot{\gamma} > 100^{1/s}$.

3.2 Material Model Fit for Squeeze Flow Simulations

3.2.1 Adhesive

The experimental investigation of the adhesive in Sect. 3.1.1 has shown that the adhesive is a yield stress material and therefore the Herschel-Bulkley-Papanastasiou model should be used to describe the data. Figure 8a shows the least squares fit, taking into account all measured viscosity data points. The model cannot describe the bump in the shear viscosity curve at $0.1^{1/s}$. Moreover, in the range of higher shear rates, the model deviates visibly from the experimental data. Especially, the range of higher shear rates becomes more important for decreasing gap heights and should therefore be better represented by the material model. In a second fit, only shear viscosity data for shear rates $> 0.1^{1/s}$, indicated by filled symbols, are taken into account for the material model fitting, cf. Fig. 8b. For this second fit, the initial yield point value was adapted to the second plateau of the stress curve. It was possible to describe the shear viscosity curve more precisely. This could be favorable for force prediction in small squeeze-flow gaps and has to be studied further in future studies. Table 1 summarizes the fitting parameters of both fits.

Table 1. Fit parameter of the Herschel-Bulkley-Papanastasiou material model.

Material	K	n	τ_γ	m	Residual	Comment
Adhesive	6743.97	0.36	400	1e6	0.983	all shear rates considered
Adhesive	335.39	0.75	8229	1e6	0.996	$\dot{\gamma} > 0.1^{1/s}$ considered

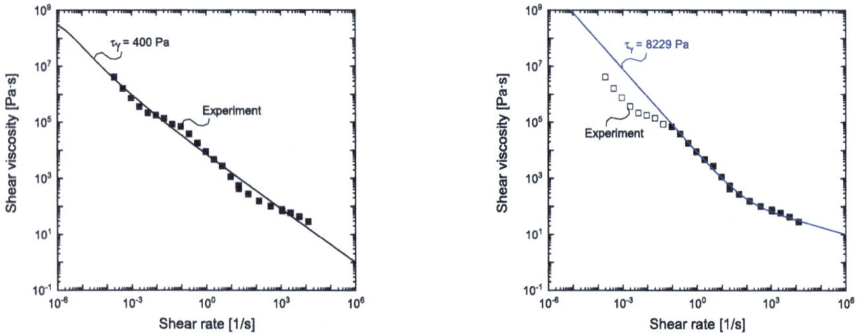

(a) All viscosity points taken into account. (b) Filled points taken into account.

Fig. 8. Herschel-Bulkley-Papanastasiou material model fit for the shear viscosity curve of the adhesive.

3.2.2 Thermal Gap Filler

As the thermal gap filler is not a yield stress material, which is indicated by the missing plateau of the shear stress for low shear rates, the Carreau model is used as a material model. Figure 9 shows the Carreau fit to the experimental viscosity data. For the fitting parameters, cf. Table 2. It was assumed that there is a zero shear rate plateau at shear rates smaller than $10^{-61/s}$.

Table 2. Fit parameter of the Carreau material model.

Material	η_0	η_{inf}	λ	n	Residual
Thermal gap filler	$7.15 \cdot 10^7$	25	$8.62 \cdot 10^5$	0.12	0.976

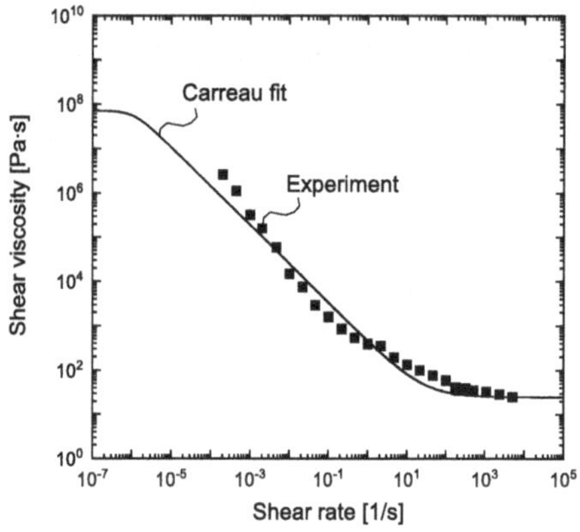

Fig. 9. Carreau fit of experimental viscosity data for the thermal gap filler.

4 Conclusion

In this study a comprehensive guideline was provided for conducting rheological measurements intended for material modeling in squeeze flow simulations. Through systematic experimental investigations using parallel plate and capillary rheometry, the flow behavior of two highly filled two component materials was analyzed. In accordance with previous findings, it was demonstrated that oscillatory measurements are unsuitable for determining the shear viscosity curve of filled materials. Instead, rotational rheometry is the preferred technique. Highly filled materials are likely to exhibit wall slip, which was observed in one material. Its influence on the shear viscosity curve was analyzed. Wall slip leads to an apparent reduction in viscosity, which can result in an underestimation of joining forces in simulations. Methods for identifying and mitigating wall slip were also investigated. One of the investigated materials showed yield stress behavior, making the Herschel-Bulkley-Papanastasiou model suitable, while the thermal gap filler followed a zero-shear viscosity profile, best described by the Carreau model. The findings emphasize the importance of selecting appropriate rheological models and test conditions that closely replicate the actual joining process. These insights not only enhance the understanding of adhesive flow during assembly but also support the development of more reliable simulation tools for optimizing adhesive application and minimizing material waste. Future work should focus on the impact of rheology data on squeeze flow simulations, parameter studies, and the comparison of different discretization methods.

References

Aparecida Silva, L., Espinosa, C., Paroissien, E., Lachaud, F., Silva, L.F.: Numerical simulations of adhesive spreading during bonding-induced squeeze. J. Adhes. **98**(16), 2517–2549 (2022)

Barnes, H.A.: A review of the slip (wall depletion) of polymer solutions, emulsions and particle suspensions in viscometers: its cause, character, and cure. J. Nonnewton. Fluid Mech. **56**(3), 221–251 (1995)

Barnes, H.A.: A review of the rheology of filled viscoelastic systems. Rheol. Rev. 1–36 (2003)

Carreau, P.J.: Rheological equations from molecular network theories. Trans. Soc. Rheol. **16**(1), 99–127 (1972)

Coussot, P.: Yield stress fluid flows: a review of experimental data. J. Nonnewton. Fluid Mech. **211**, 31–49 (2014)

Doraiswamy, D., Mujumdar, A., Tsao, I., Beris, A., Danforth, S., Metzner, A.: The cox-merz rule extended: a rheological model for concentrated suspensions and other materials with a yield stress. J. Rheol. **35**(4), 647–685 (1991)

Da Silva, L.F., Öchsner, A., Adams, R.D.: Handbook of Adhesion Technology. Springer, Cham (2011)

Engmann, J., Servais, C., Burbidge, A.S.: Squeeze flow theory and applications to rheometry: a review. J. Nonnewton. Fluid Mech. **132**(1–3), 1–27 (2005)

Genovese, D.B.: Shear rheology of hard-sphere, dispersed, and aggregated suspensions, and filler-matrix composites. Adv. Coll. Interface. Sci. **171**, 1–16 (2012)

Gleissle, W., Hochstein, B.: Validity of the cox-merz rule for concentrated suspensions. J. Rheol. **47**(4), 897–910 (2003)

Herschel, W.H., Bulkley, R.: Konsistenzmessungen von gummi-benzollösungen. Kolloid-Zeitschrift **39**, 291–300 (1926)

Hicks, C.R., Carlson, B.E., Mallick, P.: Rheological study of automotive adhesives: influence of storage time, temperature and shear rate on viscosity at dispensing. Int. J. Adhes. Adhes. **63**, 108–116 (2015)

Huf, M., Vallée, T., Fricke, H.: Numerical modelling and experimental validation of squeezing flows in the automobile production. Int. J. Adhes. Adhes. **130**, 103616 (2024)

Kaufmann, M., et al.: Experimental validation of a compression flow model of non-newtonian adhesives. J. Adhes. **98**(14), 2295–2324 (2022)

Kaufmann, M., Flaig, F., Müller, M., Fricke, H., Vallée, T.: How adhesives flow during joining. Int. J. Adhes. Adhes. **122**, 103315 (2023)

Kaufmann, M., Flaig, F., Müller, M., Fricke, H., Vallée, T.: Optimized adhesive application. Int. J. Adhes. Adhes. **130**, 103620 (2024)

Kamar, N., Hill, J.: Rheology of automotive adhesives during dispensing: a direct comparison of epoxy and rubber based systems via drag and pressure driven flows. Int. J. Adhes. Adhes. **133**, 103754 (2024)

Kovačević, V., Lučić, S., Hace, D., Glasnović, A., Šmit, I., Bravar, M.: Investigation of the influence of fillers on the properties of poly (vinyl acetate) adhesives. J. Adhes. **47**(1–3), 201–215 (1994)

Lemke, T., Bagusat, F., Köhnke, K., Husemann, K., Mögel, H.-J.: Time dependent viscosity of concentrated alumina suspensions. Colloids Surf. A **150**(1–3), 283–287 (1999)

Li, C., Mazich, K., Dickie, R.: A survey of rheological properties of one-component epoxy adhesives. J. Adhes. **32**(2–3), 127–140 (1990)

Meeten, G.H.: Effects of plate roughness in squeeze-flow rheometry. J. Nonnewton. Fluid Mech. **124**(1–3), 51–60 (2004)
Mooney, M.: Explicit formulas for slip and fluidity. J. Rheol. **2**(2), 210–222 (1931)
Müller, M., Willenbrock, S., Stahl, L., Vallée, T., Fricke, H.: Towards the efficient modelling of trapped air pockets during squeeze flow. Exp. Comput. Multiphase Flow **5**(1), 29–52 (2023)
Ovarlez, G., Cohen-Addad, S., Krishan, K., Goyon, J., Coussot, P.: On the existence of a simple yield stress fluid behavior. J. Nonnewton. Fluid Mech. **193**, 68–79 (2013)
Ong, E.E.S., O'Byrne, S., Liow, J.L.: Yield stress measurement of a thixotropic colloid. Rheol. Acta **58**, 383–401 (2019). https://doi.org/10.1007/s00397-019-01154-y
Papanastasiou, T.C.: Flows of materials with yield. J. Rheol. **31**(5), 385–404 (1987)
Rueda, M.M., et al.: Rheology and applications of highly filled polymers: a review of current understanding. Prog. Polym. Sci. **66**, 22–53 (2017)
Sanghvi, M.R., Tambare, O.H., More, A.P.: Performance of various fillers in adhesives applications: a review. Polym. Bull. **79**(12), 10491–10553 (2022)
Waele, A.D.: The manifestation of interfacial forces in dispersed systems. J. Am. Chem. Soc. **48**(11), 2760–2776 (1926)
Wei, Y., Jin, X., Luo, Q., Li, Q., Sun, G.: Adhesively bonded joints - a review on design, manufacturing, experiments, modeling and challenges. Compos. B Eng. **276**, 111225 (2024). https://doi.org/10.1016/j.compositesb.2024.111225
Yoshimura, A., Prud'homme, R.K.: Wall slip corrections for couette and parallel disk viscometers. J. Rheol. **32**(1), 53–67 (1988)

Joining Adhesive Bonds - Part 2: Understanding the Effects on Numerical Squeeze Flow Prediction

Franziska Hagemann(✉) and Manuel Schiel

Henkel AG & Co. KGaA, Henkelstraße 67, 40589 Düsseldorf, Germany
{franziska.hagemann,manuel.schiel}@henkel.com

Abstract. Joining forces play a crucial role during the assembly of adhesively bonded products. Especially in the battery production, where highly viscous thermal interface materials bridge the gap between battery cells and cooling plate, the joining force is a critical parameter. If the latter is too high, the performance of the comparatively fragile cells might be reduced, or the cell could even be damaged. Different simulation tools as well as numerical methods, either commercially available or published in scientific literature, can be used to describe the joining process mathematically. The aim of this study is, to evaluate and compare computational fluid dynamics and coupled Eulerian-Lagrangian methods regarding their applicability in modeling of joining processes of different thermal interface and adhesive materials. Different material models for the description of the shear rate-dependent viscosity are compared evaluating their contribution towards the accuracy of the simulation. This study lays a foundation to successfully integrate squeeze flow simulations into digitalized industrial work-flows by indicating how a combination of simulation approach and material model reliably predicts joining forces. Thereby, the trade-off between an accurate physical representation, experimental and numerical effort and precision is taken into account.

Keywords: Adhesive · Squeeze Flow · Rheology · Numerical Simulation

1 Introduction

Adhesive bonds are crucial in various applications. A current field of use is the automotive industry which is transitioning from internal combustion engines to battery electric drives. This requires, e.g., an efficient thermal management of the battery cells to prevent lifetime losses and thermal runaways. To thermally connect the battery cells with the cooling system, thermal interface materials (TIM) are used. Another type of adhesive material used in battery packs are polyurethane (PU) based sealings. Adhesives and TIM have complex non-Newtonian flow behavior, resulting from the polymers, fillers, and additives

L. F. M. da Silva and R. D. Adams (Eds.): AB 2025, PEM, pp. 118–137, 2026.
https://doi.org/10.1007/978-3-032-19022-2_8

which are incorporated to reach the desired adhesive's properties (Kamar and Hill (2024); Jojibabu et al. (2016)).

To assemble an adhesive joint, the liquid adhesive is dispensed and subsequently, the parts are joined. During bonding, a controlled material spreading should take place to ensure excess volume reduction and to avoid air enclosures. Moreover, the joining forces are of interest to make sure that the adherent parts do not deform. A model based engineering approach can be used to access the joining process. Modeling squeeze flow offers the possibility to improve understanding, reduce material usage and enables parameter studies. The adhesive's rheology is an important basis to model squeeze flow, as well as the contact angle between surface and adhesive. However, for high capillary number, viscous forces are dominating and the contact angle is negligible ($\mathrm{Ca} = \eta \cdot v/\sigma > 10^{-5}$, where η is the dynamic viscosity, v the velocity and σ the surface tension (Shi et al. (2018))). Kamar and Hill (2024) assessed the rheology needed for adhesive dispensing applications. Hagemann et al. (2025) investigated the rheology measurements needed for the simulation of squeeze flow.

Squeeze flow calculations are addressed in various publications and can be divided into analytical and numerical approaches. Analytical solution offer fast but less precise results, numerical approaches cost more time but can solve more complex set-ups. Analytical solutions for squeeze flow problems with viscoplastic Herschel-Bulkley, Bingham fluids and yield slip were developed by Muravleva (2013; 2019; 2020). Fusi and Ballotti (2021) solved squeeze flow equations for stress power law fluids.

Numerically, different approaches can be used to solve the joining applications: Apart from the Eulerian description which is used in computational fluid dynamic (CFD), other approaches to solve squeeze flow models numerically are Lagrangian or arbitrary Lagrangian Eulerian (ALE). A comprehensive overview of numerical simulations of the joining process is given by Aparecida Silva et al. (2022) and Engmann et al. (2005). The smoothed particle hydrodynamics (SPH) method is a powerful method to access large deformations of fluids and can overcome the meshing challenge for multiphase interfaces (Aparecida Silva et al. (2022)). However, the results of squeeze flow calculations drastically depend on the particle size and small particle sizes lead to higher simulation times. Moreover, results from simulations with non-Newtonian material models as well as the representation of the free surface still lack precision (Aparecida Silva et al. (2022); Huf et al. (2024); Breinlinger et al. (2013)). Aparecida Silva et al. (2022) studied squeeze flow with an ALE approach, describing the fluid adhesive by a Johnson-Cook solid behavior law and came to the conclusion that also this approach needs improvement. So far, Eulerian approaches offer the most promising results: A finite element analysis of squeeze flow with a Herschel-Bulkley material was developed by Adams et al. (1997) and Sherwood and Durban (1998). Up to date CFD simulations are most commonly used to simulate squeeze flow even though they face challenges of mesh size for small gaps and multiphase interfaces (Landgrebe et al. (2015); Neugebauer et al. (2012); Fricke and Vallée (2016)). To reduce simulation time of squeeze flow problems, the general partially filled gap model was developed (Müller et al. (2022; 2014; 2018; 2023); Kaufmann et al. (2022)). The main advantage of this model is the assumption that pressure is constant in the through-thickness direction which allows for using only one element layer in the gap. It has not been implemented in commercial solvers or released as open-source so far. Therefore the industrial use is limited.

Recent studies approached squeeze flow by different computational approaches, using rheological data to describe the squeezed material: Spangenberg et al. (2018) simulated squeeze flow for the assembly of wind turbine blades using CFD; Huf et al. (2024) focused on automotive applications using a coupled Euler-Lagrange and a SPH approach. These studies primarily concentrate on post-processing simulation results, often placing limited emphasis on the rheological data generation and the material model fitting. This study explores the impact of rheological parameters taken from Hagemann et al. (2025). It is the aim to study the sensitivity of parameters on the squeeze flow result. Moreover, a finite volume and a finite element method approach are compared and the influence of wall slip and different squeeze velocities is investigated. Furthermore, the Arbitrary Lagrangian Eulerian (ALE) method is compared to these methods.

Through this analysis, we aim to highlight the importance of the appropriate rheological evaluation for the presented use case. It should help to understand the complex behavior in squeeze flow applications and could be extended to predict adherent deformation or adhesive distribution in the fluid state.

2 Material and Methods

2.1 Materials

In this work two different materials initially investigated by Hagemann et al. (2025) were used. Both are used in battery electric vehicles, are two-component materials and highly filled:

- Two-component thermal gap filler: (Part A and Part B), mixing ratio 1:1. Squeeze flow experiments were conducted with the mixture.
- Two-component adhesive: Adhesive and booster, mixing ratio 40:1. Squeeze flow experiments were only conducted with the adhesive.

The Carreau model was used to describe the thermal gap filler, cf. Eq. 1. Parameters were taken from Hagemann et al. (2025) and are summarized in Table 1.

$$\eta = \eta_{\mathrm{inf}} + (\eta_0 - \eta_{\mathrm{inf}}) \cdot \left(1 + (\lambda \cdot \dot{\gamma})^2\right)^{\frac{n-1}{2}} \quad (1)$$

Table 1. Fit parameter of the Carreau material model describing the thermal gap filler.

No.	η_0 [Pa · s]	η_{inf} [Pa · s]	λ [s]	n [-]	Residual	Figure
4	$7.15 \cdot 10^7$	25	$8.62 \cdot 10^5$	0.12	0.976	4

The adhesive without the booster was described by a Herschel-Bulkley-Papanastasiou model:

$$\eta = \frac{\tau_{\dot{\gamma}}}{\dot{\gamma}} \cdot (1 - \exp(-m \cdot \dot{\gamma})) + K \cdot \left(\frac{\dot{\gamma}}{\dot{\gamma}_{\text{ref}}}\right)^{n-1} \tag{2}$$

The parameter set was taken from Hagemann et al. (2025) and is summarized in Table 2. The yield point for the material model was derived from the plateau regions of the shear rate dependent stress curve, cf. Fig. 1. An exact determination of the beginning of the plateau between 0.1 1/s and 100 1/s is difficult and slight variations of the second-plateau-yield point are possible. This is why a third parameter set with an adapted yield point is added to the table. The impact of the yield point on squeeze flow curves is discussed in Sect. 4.2.

The Herschel-Bulkley-Papanastasiou differs from the Herschel-Bulkley model by the term $\exp(-m \cdot \dot{\gamma})$ and hence requires further determination of the parameter m. The case presented here does not benefit from this additional complexity and the Herschel-Bulkley model could be used as well. Still, depending on the solver, the one or the other model is implemented. Therefore the option of using a large value for m can be used.

Table 2. Fit parameter of the Herschel-Bulkley-Papanastasiou material model describing the adhesive.

No.	K [Pa · s]	n [-]	τ_γ [Pa]	m [s]	Residual	Comment	Figure
1	6743.97	0.36	400	1e6	0.983	all shear rates considered	5a, 5b
2	335.39	0.75	8229	1e6	0.996	$\dot{\gamma} > 0.09$ 1/s considered	5c, 5d
3	335.39	0.75	6000	1e6	0.996	Reduction of yield stress	5c, 5d

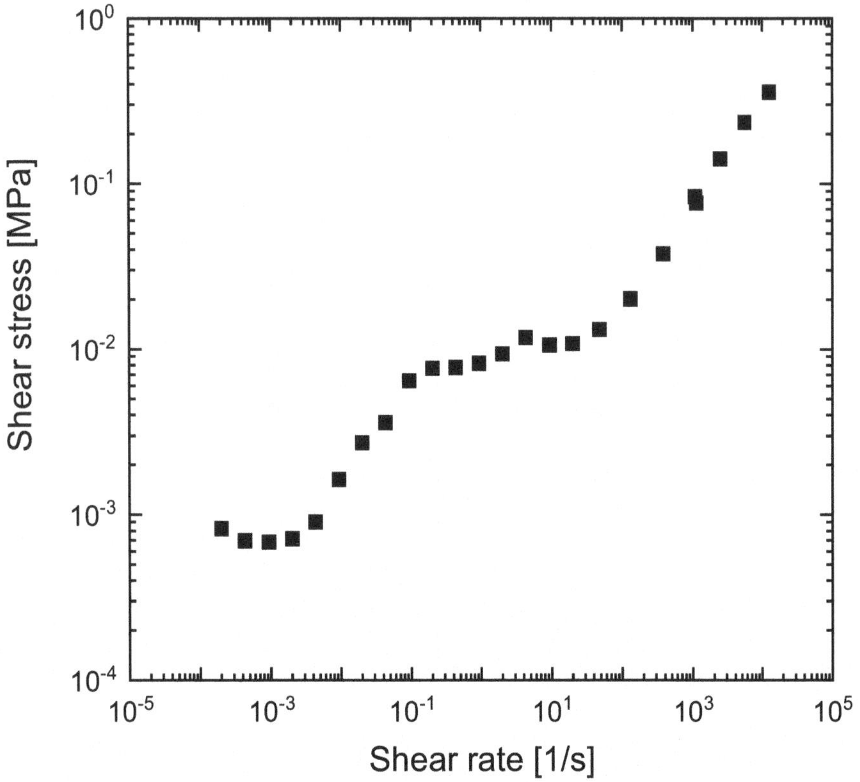

Fig. 1. Shear rate dependent stress curve for the adhesive taken from Hagemann et al. (2025).

2.2 Methods

All squeeze flow experiments were conducted at 23 °C and each experiment was repeated three times. The experiments were carried out with a 5 kN universal testing machine from Zwick Roell. The top stamp diameter was 40 mm and the bottom plate diameter was 70 mm. The material was applied to the bottom plate, so that a 5 mm gap was completely filled. The stamp was driven down to a gap height of 5 mm, stopped, and the squeezed out (extruded) material was trimmed. Subsequently, the squeeze flow test was started and the gap height was reduced to 0.5 mm. The velocity of the stamp was set to a constant value of 1 mm/s for both materials. In a second step, the squeeze velocity for the experiments with the adhesives was changed to 0.1 mm/s and 5 mm/s. A schematic representation of the squeeze flow set-up is given in Fig. 2.

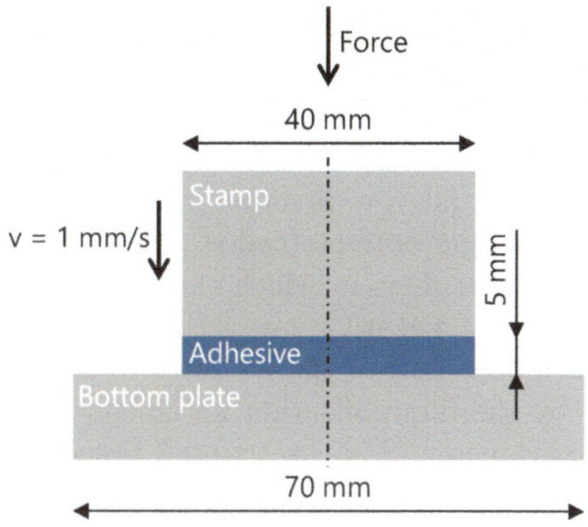

Fig. 2. Schematic of the squeeze flow set-up.

3 Modeling

Squeeze flow was simulated similar to the experimental set-up described in Sect. 2.2. Different types of fluid simulations were realized: With Comsol Multiphysics® 6.3 a finite element method was built and a finite volume method was realized with Ansys Fluent® 2024 R2. A combination of structural and fluid simulation techniques was performed using LS-Dyna R16.1 using the available structured arbitrary Lagrangian Eulerian (S-ALE) approach. Figure 3 visualizes the different model set-ups used for simulations.

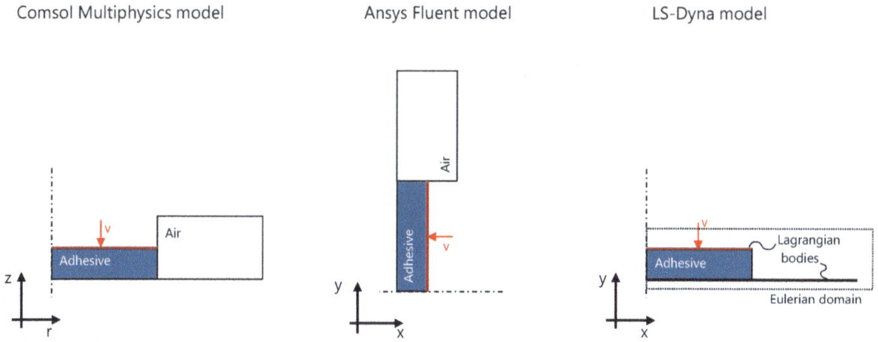

Fig. 3. Schematics of model set-ups for Comsol, Fluent and LS-Dyna.

3.1 COMSOL Multiphysics

In Comsol Multiphysics the squeeze flow model, represented in Fig. 2, was reduced to a 2D axisymmetric geometry. Materials which were added to the model were: adhesive which was located in the gap for t = 0 s, and air which was surrounding stamp and adhesive. The adhesive's density and inelastic non-Newtonian viscosity model were adapted to the respective material. A moving mesh setting was needed, including a symmetry condition and a prescribed mesh displacement of 1 mm/s in negative z direction representing the movement of the stamp. Two different modeling options (called physics in the context of COMSOL) were used: (1) Laminar flow, in which the symmetry axis and the outlet were defined, and gravity in negative z-direction was included. (2) Level set which defines the multiphase flow of the adhesive in air. The resolution of the interface was defined by the reinitialization parameter =0.5 and the interface thickness controlling parameter IS was set to 0.5 times its default value (Mirjalili et al. [2020]). In this study, these parameters increased the resolution of the interface's shape but did not impact the resulting squeeze forces. At the walls, a no-slip condition was assumed in accordance with the results of the rheological investigations. A mesh dependency study was performed to determine a robust, numerically stable, and correct solution. While 4 elements across the gap resulted in numerical instabilities and non-convergence of the solver, 20 elements across the thickness do not show any meaningful difference compared to the used configuration with 10 elements through thickness. The remaining meshing parameters are summarized in Table 3. The study consisted of two steps: The first step was the phase initialization. Subsequently, a time dependent calculation from t = 0 s to 4.5 s was performed resulting in a final gap height of 0.5 mm.

Table 3. Element size parameters of the mesh

Maximum element size	$3.5 \cdot 10^{-4}$ m
Minimum element size	$1 \cdot 10^{-5}$ m
Maximum element growth	1.13
Curvature factor	0.3
Resolution in narrow regions	1

3.2 Ansys Fluent

Also in Ansys Fluent a 2D axisymmetric geometry was used for the simulation. The symmetry axis must be the x-axis of the model. Gravity was included in negative x-direction. A multiphase model with a volume of fluid was chosen. Air was set as primary and the adhesive as secondary phase. The viscous model was set to laminar. In the boundary conditions interfaces between air and adhesive were defined as well as walls and the outlet. Moreover a mesh interface between the initial air and adhesive cell zone was defined. To simulate the downwards moving stamp a dynamic mash was added to the model. Mesh methods were set to smoothing. All dynamic mesh zones were set as deforming mesh type, except for the stamp surface which is a rigid body. A user defined motion function that describes the constant velocity of 1 mm/s in negative x-direction was added. After performing a mesh independence study, simulations were run with the following settings: An all quadrilateral mesh with a mesh size of 0.01 mm was used. The step size in the simulation was set to 0.001 s.

3.3 LS-Dyna

In LS-Dyna the structured arbitrary Lagrangian Eulerian (S-ALE) method was used to create a coupled problem. The method offers the possibility to extend this study towards including effects resulting from fluid-structure-interaction and is therefore of high relevance. The S-ALE method preferred here over the classical arbitrary Lagrangian Eulerian (ALE) method as huge savings in simulation time can be reached Chen (2020). The main difference between both methods is the structured mesh of the Eulerian domain in S-ALE vs. ALE. This results in numerical advantages.

To further reduce simulation time, a two-dimensional axisymmetric approach was used. The two plates were modeled as Lagrangian rigid bodies, meaning that these plates do not allow for deformation. The Eulerian mesh is configured with the dimensions of 50 mm $\times$ 6 mm $\times$ 1 mm. Although creating an axisymmetric model, it is still necessary to define one element in the out-of-plane direction due to implementation reasons in the solver, meaning a three dimensional Eulerian domain is surrounding both two-dimensional plates.

A homogeneous mesh was created using *ALE_STRUCTURED_MESH_CONTROL_POINTS modeling 150 elements in x- and 36 elements in y-direction. The volume between the top and the bottom plate was filled with the adhesive using *ALE_STRUCTURED_MESH_VOLUME_FILLING. The remaining volume of the Eulerian domain is essentially void (*MAT_VACUUM). Fluid-structure-interaction is activated using the keyword *ALE_STRUCTURED_FSI. With that, the Lagrangian bodies are coupled with the adhesive material of the respective material group (*SET_MULTI_MATERIAL_GROUP_LIST). The boundary conditions were applied to the two rigid bodies using *BOUNDARY_SPC_SET and the velocity was defined using *BOUNDARY_PRE-SCRIBED_MOTION_RIGID in combination with a smooth velocity curve (*DEFINE_CURVE_SMOOTH) to avoid inertia effects during acceleration.

4 Results and Discussion

In this section the impact of viscosity data parametrization on squeeze flow simulation results is discussed. The required material models for the non-Newtonian fluids are taken from Hagemann et al. (2025). In the first part, the results for the Carreau fluid are presented and in the second part a profound investigation of the Herschel-Bulkley fluid is presented. The latter focuses on the impact of the yield stress and squeeze velocity, the used numerical simulation method, and the effect of wall slip. A parametric study and the non-existing impact of the contact angle are presented and discussed in the supplementary information, cf. Sect. 5.2. The contact angle has no impact on the squeeze flow forces, because the viscosity of the adhesive is the dominating factor.

4.1 Thermal Gap Filler: Carreau Fluid

The thermal gap filler has no yield point and is therefore described by the Carreau material model (Hagemann et al. 2025)). Figure 4a shows the Carreau fit of the experimental viscosity data. The fitting parameters are summarized in Table 1. It was assumed that a zero shear rate plateau exists at shear rates outside of the investigated shear rate range ($<10^{-6}$ 1/s). The results of the corresponding squeeze flow simulation are presented in Fig. 4b. The squeeze flow prediction of the Carreau fluid is considered accurate, even though the viscosity fit shows deviations from the experimental data. These differences, especially at small gap height, can well be related to measurement accuracy as global displacement was taken into account. To improve experimental robustness a more complex experimental setup as presented in Kaufmann et al. (2022), will be needed.

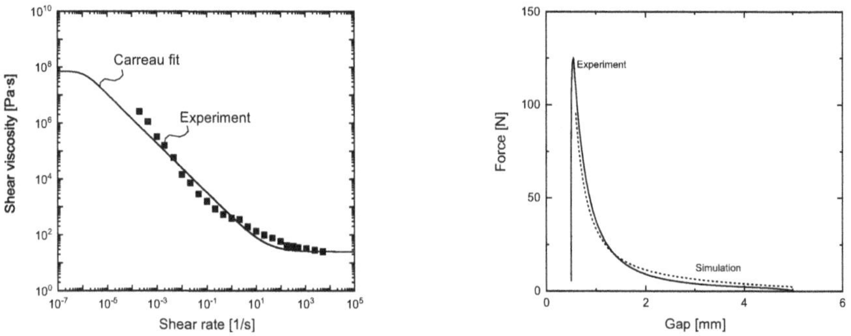

(a) Carreau fit of viscosity data.

(b) Squeeze flow validation.

Fig. 4. Material model fit and squeeze flow validation of the thermal gap filler. Viscosity data and parameter fits were taken from Hagemann et al. (2025). Squeeze flow simulation was performed with Comsol Multiphysics.

4.2 Adhesive: A Herschel-Bulkley Fluid

In this part, the results for the yield point adhesive are discussed, taking a deeper look into different impact factors.

4.2.1 The Impact of Yield Point and Squeeze Velocity

Figure 5 compares different material model fits based on the same experimental viscosity curve and presents the corresponding squeeze flow simulation results alongside experimental data.

In Fig. 5a the full viscosity dataset was used to determine the parameters (parameter set No. 1 in Table 2). However, the resulting squeeze-flow simulation (Fig. 5b) shows significant deviations from the experimental data. This discrepancy arises because the model does not accurately capture the viscosity at high shear rates, and the yield point is poorly defined. A detailed parameter study (see supplementary information) indicates that an accurate representation of high shear rate behavior is crucial for reliable squeeze-flow predictions. The shear stress curve in Fig. 1 exhibits two plateau regions. If the first plateau is interpreted as the yield stress, only the low-shear region of the viscosity curve (up to 0.01 1/s, shown in Fig. 5a) can be fitted well. However, this region has less influence on the squeeze flow behavior than the high-shear region. Therefore, in Fig. 5c, only data above 0.1 1/s were considered for fitting.

In this case, the yield stress was adjusted to correspond to the second plateau in the shear stress curve. This resulted in a significantly improved fit using parameter set No. 2 (Table 2). With these parameters, the HerschelBulkley-Papanastasiou model captures the high shear rate viscosity more accurately, leading to better agreement between simulation and experiment (Fig. 5d). Further reducing the yield stress parameter τ_γ to 6000 Pa (parameter set No. 3 in Table 2) still yields a value consistent with the second plateau in Fig. 1 and the viscosity curve remains well described (Fig. 5c). As the squeeze flow simulation is highly sensitive to this parameter, the agreement with experimental data improves further, as shown in Fig. 5d.

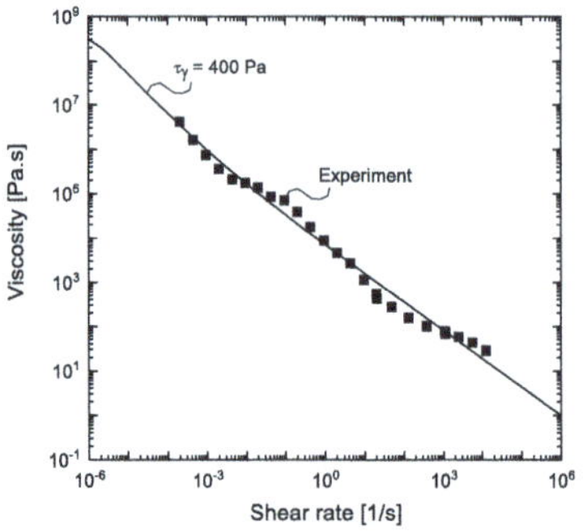

(a) Herschel-Bulkley-Papanastasiou fit. All data points were considered, cf. No. 1.

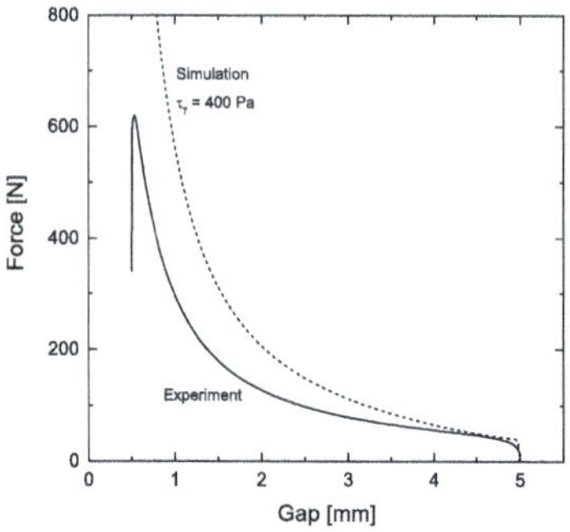

(b) Comparison of experimental and simulation results for squeeze flow. Fit for all data points was used, cf. No. 1.

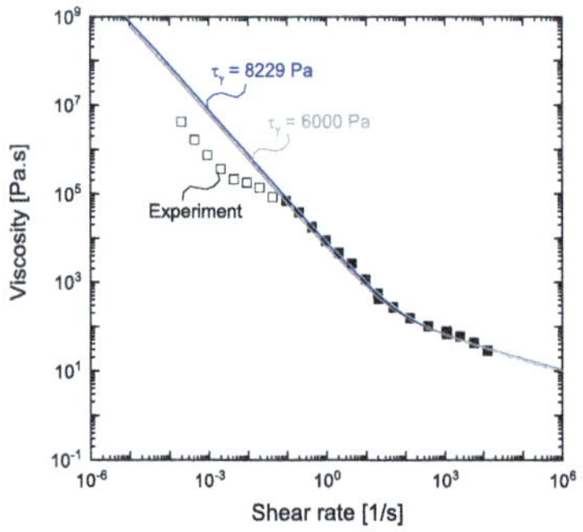

(c) Herschel-Bulkley-Papanastasiou fit. Filled data points were considered, cf. No. 2 and 3.

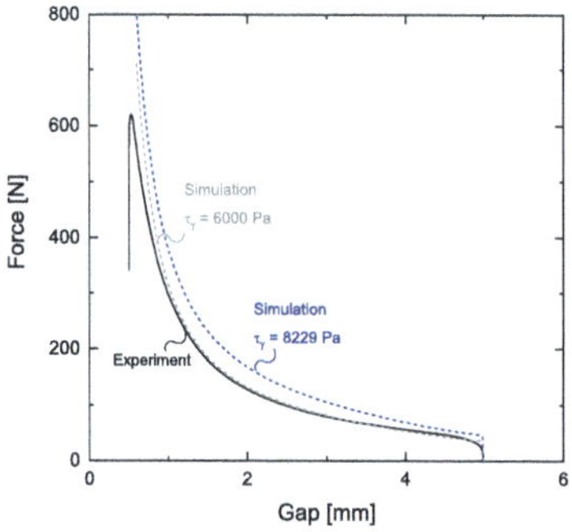

(d) Comparison of experimental and simulation results for squeeze flow. Parameter-sets No. 2 and 3 were used.

Fig. 5. Material model fit and squeeze flow validation of the adhesive. Viscosity data and parameter fits were taken from Hagemann et al. (2025). Squeeze flow simulation was performed with Comsol Multiphysics. Parameter-sets are taken from Table 2.

The results show that a good agreement of experimental data and material model at high shear rates, especially where the shear viscosity runs into a plateau, is important for the simulation of squeeze flow: The smaller the gap between stamp and bottom plate gets, the higher the velocity and the shear rates, and the higher the force that is needed to keep the squeeze velocity constant. In the following, the parameter-set No. 3 from Table 2 will be used.

In Fig. 6 the experimental and simulation results for different squeeze velocities are shown for two different yield stresses. The parameter set No. 3 from Table 2 was used for Fig. 6a. The simulations for a squeeze velocity of 0.1 mm/s, 1 mm/s, and 5 mm/s describe the experimental data with sufficient accuracy. In Fig. 6b, the yield point was increased to 7000 Pa·s. This new yield point parameter is introduced for better visibility. For 5 mm/s and 1 mm/s the simulation results are still accurate. However, for 0.1 mm/s, the prediction deviates from the experiment. An assumption for this behavior is, that for a squeeze velocity of 0.1 mm/s the shear stress distribution in the gap is lower and just high enough, that the material flows. For higher velocities this effect becomes less important, as the shear stress level in the material is higher. Furthermore, a potential thixotropic behavior of the material cannot be represented in the Herschel-Bulkley material model. This time and rate dependent effect is to be investigated further in future studies. This variation of the squeeze velocity emphasizes the importance of a well defined and most precise material model as input for the simulation. Moreover, validation experiments are important to prove the prediction as the simulation is very sensitive towards Herschel-Bulkley material model parameters.

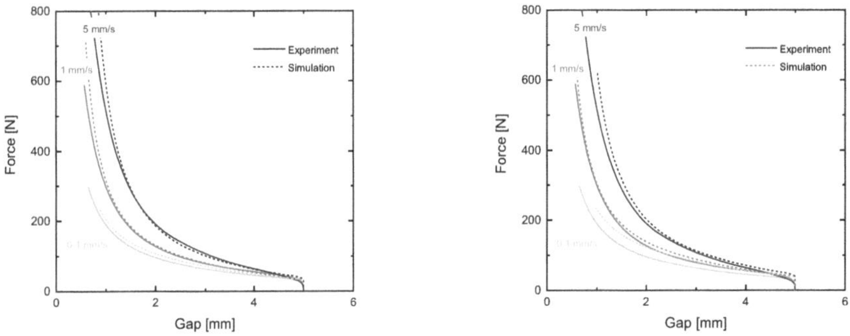

(a) Yield stress: $\tau_{\dot{\gamma}}$ = 6000 Pa·s (b) Yield stress: $\tau_{\dot{\gamma}}$ = 7000 Pa·s

Fig. 6. Comparison of experiment and COMSOL CFD simulation for different squeeze velocities (5 mm/s, 1 mm/s, and 0.1 mm/s).

4.2.2 Impact of Numerical Simulation Methods

In Fig. 7 force versus gap height curves from different solvers are compared. The result of the finite volume approach used by Ansys Fluent, differs slightly from one obtained with the finite differences approach used by Comsol Multiphysics. As both simulations were investigated regarding mesh independence, a reason for the deviation between both results might be based on numerical differences or the fact that Ansys Fluent uses the Herschel-Bulkley and Comsol the Herschel-Bulkley-Papanastasiou model. Compared to the CFD based solutions, the ALE approach in LS-Dyna significantly underestimates the reaction force especially at large gaps. Also the characteristic stepwise increase of the force at the beginning (at 5 mm) is not represented. Both effects could be related to a different evaluation of shear forces within the material or at the Lagrangian-Eulerian interface. In terms of computational effort it can also be said that although in all three cases axisymmetric 2D models were used, the ALE approach takes significantly longer to complete. At 10 cores, approximately 12 h are needed on state-of-the art computational infrastructure. For the CFD approaches calculation times below 1 h were needed. This in turn means that realistic three-dimensional cases which might also include more complex behavior of the Lagrangian bodies is hard or even impossible to address.

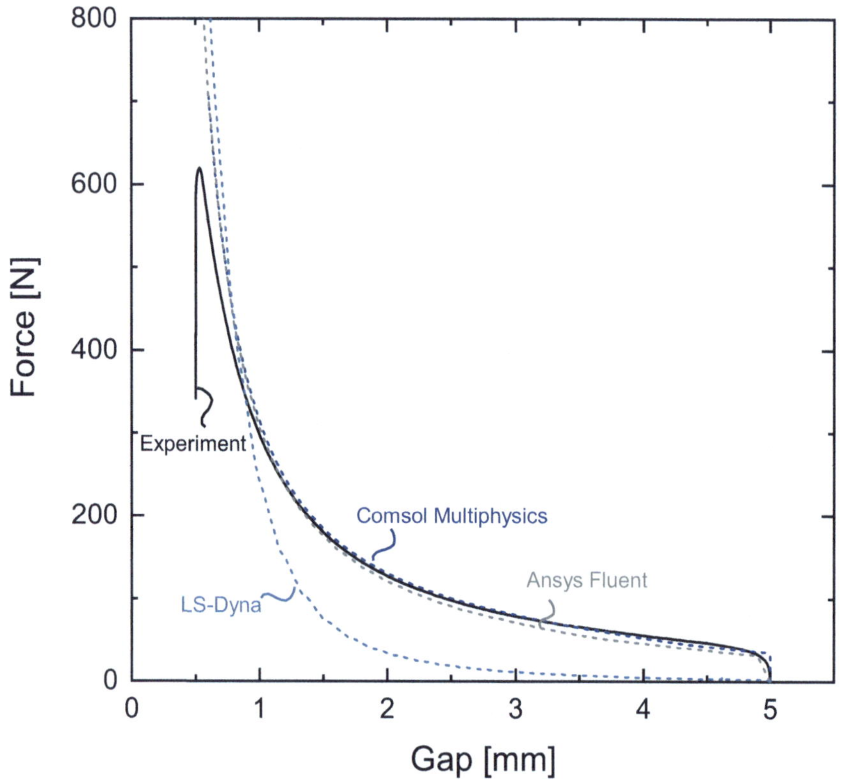

Fig. 7. Comparison of different CFD software: The blue dashed line represents the results obtained with Comsol Multiphysics, the gray dashed line is calculated with Ansys Fluent, and the light blue dashed line with LS-Dyna. The simulation results are compared to experimental data for the adhesive using the Herschel-Bulkley-Papanastasiou model in Comsol and the Herschel Bulkley model in Ansys Fluent and LS-Dyna.

4.2.3 Looking Inside: Velocity, Shear Rate, and Viscosity Distribution

Velocity, shear rate and viscosity distribution in the gap for the adhesive are shown in Fig. 8. These images result from squeeze flow simulations with the parameter-set No. 3. As no-slip conditions at the walls were assumed, the velocity at the boundary layer is zero, increasing to the center of the gap. The shear rate distribution is inverse to that: highest at the boundary layer and smallest in the center of the gap.

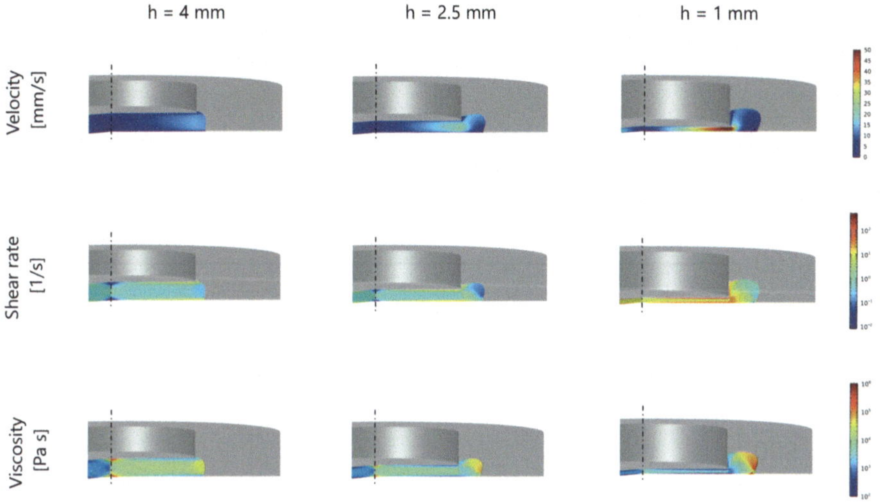

Fig. 8. Simulation results of the adhesive simulation with Comsol Multiphysics. Velocity, shear rate and viscosity profiles are shown for different gap heights.

4.2.4 Impact of Slip

It was experimentally shown that, for the one of the two materials discussed in this contribution, wall slip can occur depending on the surface roughness (Hagemann et al. (2025)). Therefore, the impact of wall slip is further discussed in this section.

To visualize the impact of wall slip on squeeze flow results, an imaginary wall slip of $\beta = 5\,\text{mm}$ was assumed, cf. Fig. 9.

Figure 10 shows the impact of wall slip on the joining forces. If the material shows slip, the flow resistance at the walls is reduced and less force is needed to squeeze the corresponding material. This finding suggest that the adherent's surface is of great importance to determine the required force.

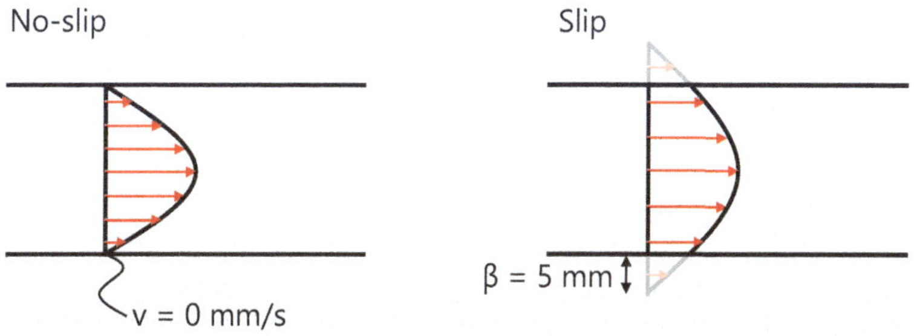

Fig. 9. Definition of no-slip and slip conditions at walls.

This statement may appear to contradict the conclusions reported by Kaufmann etal. (2023), who found that surface treatment does not influence the joining force. However, this discrepancy can be explained by the fact that different materials were investigated in the respective studies. Kaufmann etal. were unable to determine conclusively whether wall slip occurred in their experiments with silicon oil. Consequently, it is possible that a no-slip condition was already present, rendering the applied surface treatment ineffective.

Therefore, it is of even higher importance to perform viscosity measurements using a similar surface quality and to quantify the amount of wall slip and parametrize it accordingly in the numerical simulations.

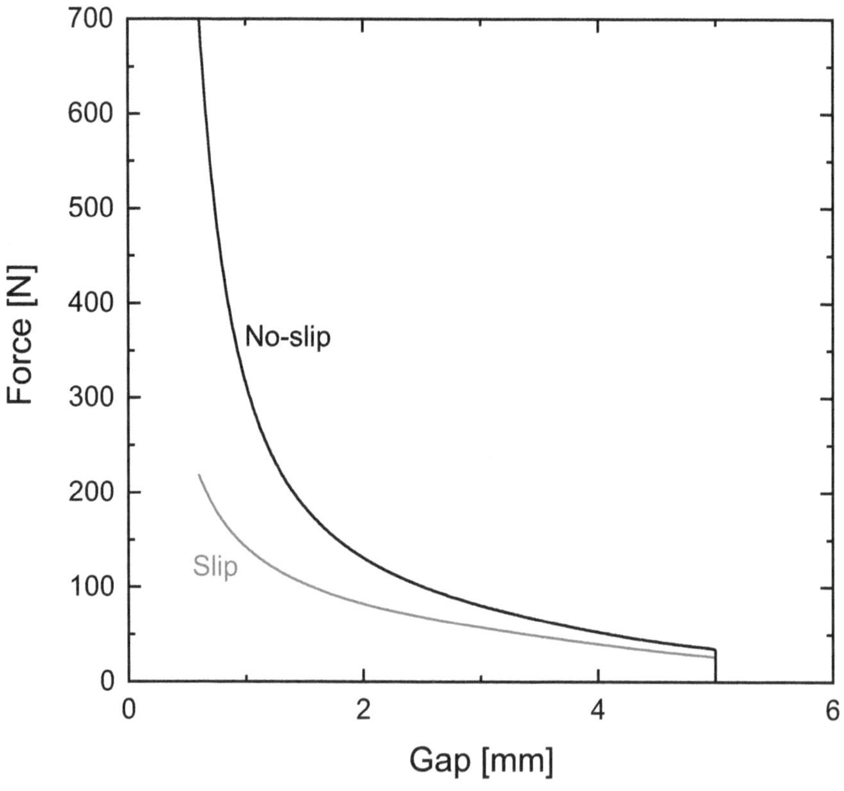

Fig. 10. Impact of wall-slip on squeeze flow forces using COMSOL simulations.

5 Conclusion

This study evaluates the predictive capabilities of numerical simulation methods for modeling squeeze flow in adhesive bonding processes. The focus is on the squeeze flow behavior of non-Newtonian materials which can be described by the Herschel-Bulkley and the Carreau material model, taken from Hagemann et al. (2025). Three numerical approaches, represented by the different softwares COMSOL Multiphysics, Ansys Fluent and LS-Dyna are compared regarding their ability to predict squeeze flow of non-Newtonian materials.

The findings underscore the critical importance of accurate shear viscosity modeling, particularly for yield-stress fluids. The sensitivity of squeeze flow simulations to the yield stress parameter highlights the necessity of precise rheological characterization. Furthermore, the study demonstrates that wall slip conditions and surface interactions significantly affect the predicted joining forces, emphasizing the need for surface-specific calibration in simulation workflows.

The Carreau model exhibited robust predictive performance, suggesting that materials without yield stress may be more reliably simulated under standard conditions. Overall, this work provides a validated framework for integrating squeeze flow simulations into digitalized adhesive bonding processes, enabling improved process design, material selection, and quality assurance in industrial applications.

Supplementary Information

5.1 Parameter Investigation

The results shown in Fig. 5 suggest to perform a parameter study for the parameters of the Herschel-Bulkley-Papanastasiou model. Both, the calculated viscosity from Eq. 2 and the resulting force function from the squeeze flow simulation in Comsol Multiphyiscs are compared in Figs. 11 and 12. The parameters K and n define the viscosity plateau region at high shear rates, cf. Fig. 11a and 11c, respectively.

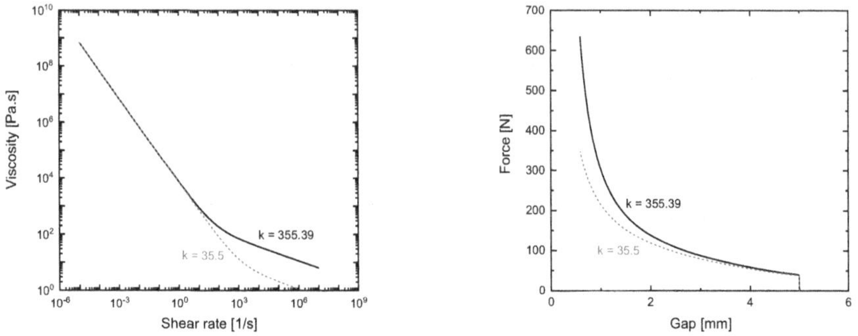

(a) Viscosity curve: Variation of K.

(b) Squeeze curve: Variation of K.

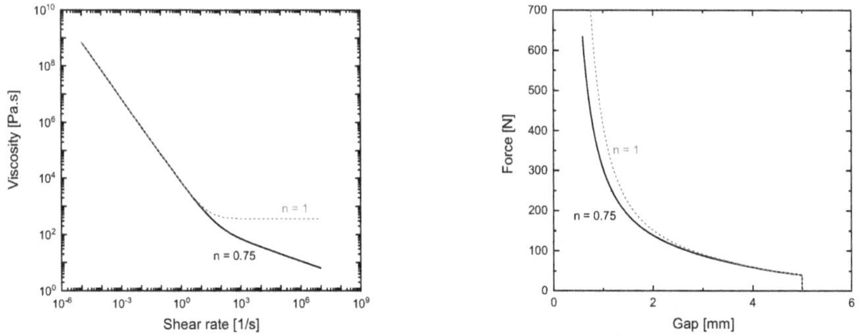

(c) Viscosity curve: Variation of n.

(d) Squeeze curve: Variation of n.

Fig. 11. Impact of Herschel-Bulkley parameter K and n on the viscosity curve (left) and on the squeeze flow curve (right). The base line parameters for the black line are: K = 355.39 Pa · s, n = 0.75, τ = 7000 Pa.s, m = 1e6 s. In the different graphs, the impact of one parameter is investigated, respectively.

In Fig. 12 the impact of the yield stress τ and the numerical smoothing parameter m on the viscosity and the squeeze flow curve is evaluated. While a reduction of the yield point shifts the viscosity curve to smaller shear rates (cf. Fig. 12a) it decreases the force drastically (cf. Fig. 12b). The parameter m has no impact on the squeeze flow curve if assumed sufficiently high. It can be concluded, that the validation of the validation of the squeeze flow simulation with experimental data is highly important, because even small variations in the yield point parameter can cause major deviations in the predicted curves.

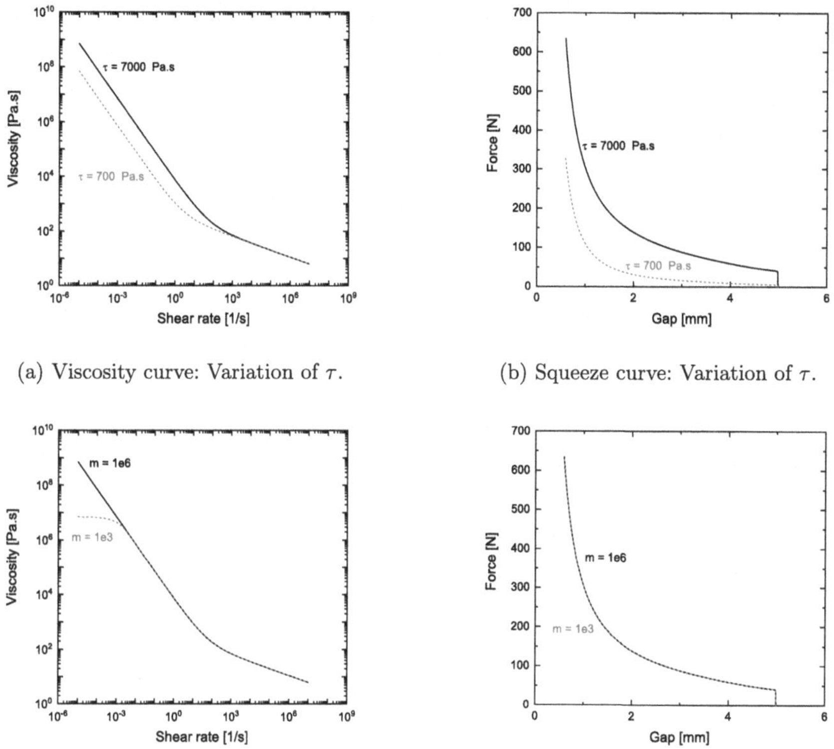

(a) Viscosity curve: Variation of τ.

(b) Squeeze curve: Variation of τ.

(c) Viscosity curve: Variation of m.

(d) Squeeze curve: Variation of m.

Fig. 12. Impact of Herschel-Bulkley parameters τ and m on the viscosity curve (left) and on the squeeze flow curve (right). The base line parameters for the black line are: $K = 355.39\,\text{Pa} \cdot \text{s}$, $n = 0.75$, $\tau = 7000\,\text{Pa.s}$, $m = 1e6\,\text{s}$. In the different graphs, the impact of one parameter is investigated, respectively.

5.2 Impact Contact Angle

In literature the contact angle between surface and adhesive is stated as an important parameter for modeling squeeze flow. However, for the investigated material Fig. 13 shows, that the contact angle has no impact on the squeeze flow forces. The reason for this are the dominating viscous forces resulting in a high capillary number.

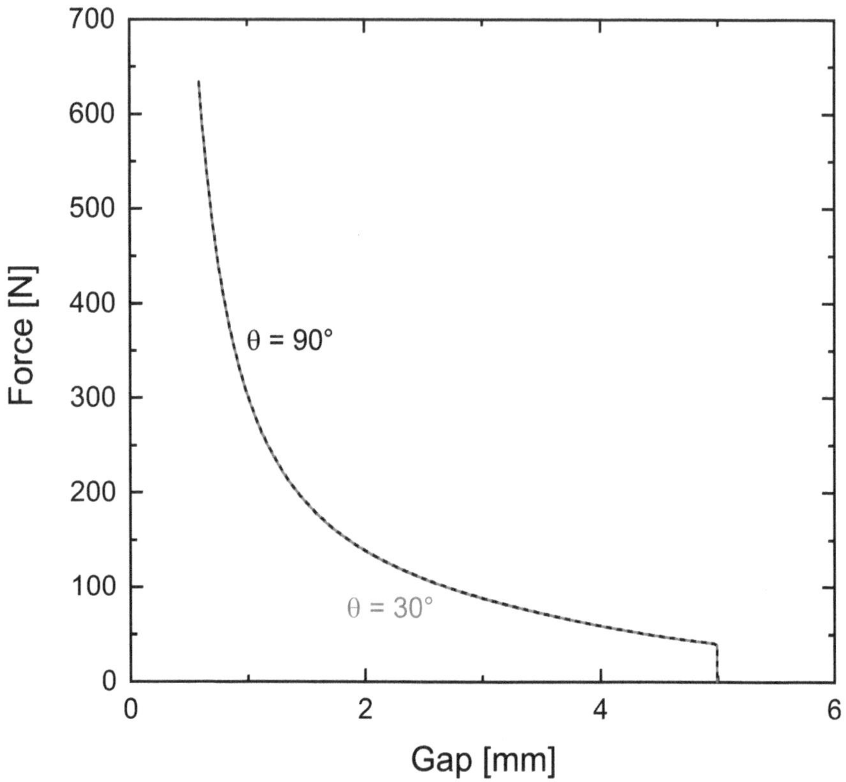

Fig. 13. The contact angle has no impact on the squeeze flow curve, because the viscosity of the adhesive is the dominating material property.

References

Adams, M., Aydin, I., Briscoe, B., Sinha, S.: A finite element analysis of the squeeze flow of an elasto-viscoplastic paste material. J. Nonnewton. Fluid Mech. **71**(1–2), 41–57 (1997)

Ansys Inc: ANSYS Inc. LS-DYNA nonlinear finite element program, ANSYS Inc (2025). ANSYS LS-DYNA Keyword User's Manual, R16 edn

Aparecida Silva, L., Espinosa, C., Paroissien, E., Lachaud, F., Silva, L.F.: Numerical simulations of adhesive spreading during bonding-induced squeeze. J. Adhesion **98**(16), 2517–2549 (2022)

Breinlinger, T., Polfer, P., Hashibon, A., Kraft, T.: Surface tension and wetting effects with smoothed particle hydrodynamics. J. Comput. Phys. **243**, 14–27 (2013)

Chen, H.: Recent Developments in LS-DYNA S-ALE. In: 16th International LS-DYNA Users Conference, (2020)

Engmann, J., Servais, C., Burbidge, A.S.: Squeeze flow theory and applications to rheometry: a review. J. Nonnewton. Fluid Mech. **132**(1–3), 1–27 (2005)

Fusi, L., Ballotti, A.: Squeeze flow of stress power law fluids. Fluids **6**(6), 194 (2021)

Fricke, H., Vallée, T.: Numerical modeling of hybrid-bonded joints. J. Adhesion **92**(7–9), 652–664 (2016)

Hagemann, F., Remmler, T., Schiel, M.: (2025)

Huf, M., Vallée, T., Fricke, H.: Numerical modelling and experimental validation of squeezing flows in the automobile production. Int. J. Adhes. Adhes. **130**, 103616 (2024)

Jojibabu, P., Jagannatham, M., Haridoss, P., Ram, G.J., Deshpande, A.P., Bakshi, S.R.: Effect of different carbon nano-fillers on rheological properties and lap shear strength of epoxy adhesive joints. Compos. A Appl. Sci. Manuf. **82**, 53–64 (2016)

Kaufmann, M., et al.: Experimental validation of a compression flow model of non-newtonian adhesives. J. Adhesion **98**(14), 2295–2324 (2022)

Kaufmann, M., Flaig, F., Müller, M., Fricke, H., Vallée, T.: Do surface pretreatments for adhesives influence the squeeze flow? Int. J. Adhes. Adhes. **124**, 103362 (2023). https://doi.org/10.1016/j.ijadhadh.2023.103362

Kamar, N., Hill, J.: Rheology of automotive adhesives during dispensing: a direct comparison of epoxy and rubber based systems via drag and pressure driven flows. Int. J. Adhes. Adhes. **133**, 103754 (2024)

Landgrebe, D., et al.: Adhesive distribution and global deformation between hybrid joints. Key Eng. Mater. **651**, 1465–1471 (2015)

Mirjalili, S., Ivey, C.B., mani, A.: A conservative diffuse interface method for two-phase flows with provable boundedness properties. J. Comput. Phys. **401**, 109006 (2020)

Müller, M., Finke, J., Stahl, L., Tong, Y., Fricke, H., Vallée, T.: Development and validation of a compression flow model of non-newtonian adhesives. J. Adhesion **98**(9), 1260–1297 (2022)

Müller, M., Jäschke, H., Bubser, F., Ostermeyer, G.-P.: Simulative studies of tribological interfaces with partially filled gaps. Tribol. Int. **78**, 195–209 (2014)

Müller, M., Tong, Y., Fricke, H., Vallée, T.: An efficient numerical model for the evaluation of compression flow of high-viscosity adhesives. Int. J. Adhes. Adhes. **85**, 251–262 (2018)

Muravleva, L.: Squeeze flow of a viscoplastic bingham medium: an asymptotic solution, (2013). arXiv:1306.6174 arXiv preprint

Muravleva, L.: Axisymmetric squeeze flow of a herschel-bulkley medium. J. Nonnewton. Fluid Mech. **271**, 104147 (2019)

Muravleva, L.: Squeeze flow of bingham, casson and herschel-bulkley fluids with yield slip at the wall by accelerated augmented lagrangian method. J. Nonnewton. Fluid Mech. **282**, 104320 (2020)

Müller, M., Willenbrock, S., Stahl, L., Vallée, T., Fricke, H.: Towards the efficient modelling of trapped air pockets during squeeze flow. Exp. Comput. Multiphase Flow **5**(1), 29–52 (2023)

Neugebauer, R., Israel, M., Mayer, B., Fricke, H.: Numerical and experimental studies on the clinch-bonding and RIV-bonding process. Key Eng. Mater. **504**, 771–776 (2012)

Sherwood, J., Durban, D.: Squeeze-flow of a herschel-bulkley fluid. J. Nonnewton. Fluid Mech. **77**(1–2), 115–121 (1998)

Spangenberg, J., Uzal, A., Nielsen, M., Hattel, J.: A robustness analysis of the bonding process of joints in wind turbine blades. Int. J. Adhes. Adhes. **85**, 281–285 (2018)

Shi, Z., Zhang, Y., Liu, M., Hanaor, D.A., Gan, Y.: Dynamic contact angle hysteresis in liquid bridges. Colloids Surf. A **555**, 365–371 (2018)

Joint design

Adhesive Bonding of Cast Aluminium for Automotive Applications: A Comparative Study

Ana A. Correia[1,2], Eduardo A. S. Marques[1(✉)], Ricardo J. C. Carbas[1], Bruno Ferreira[3], and Lucas F. M. Silva[2]

[1] Instituto de Ciência e Inovação em Engenharia Mecânica e Engenharia Industrial (INEGI), Porto, Portugal
emarques@fe.up.pt
[2] Departamento de Engenharia Mecânica, Faculdade de Engenharia (FEUP), Universidade Do Porto, Porto, Portugal
[3] Wedrone, Porto, Portugal

Abstract. This study investigates the adhesive bonding performance of die-cast aluminium alloy $AlSi_{10}Mg$, focusing on its application in the automotive industry. Two adhesive systems were evaluated: a cyclic olefin-based hot melt adhesive (LSXU) and a silane-modified polymer adhesive (Teroson MS 939). Single lap joints (SLJs) were manufactured under two substrate conditions (as-cast and machined) and three surface treatments (degreased, sanded, anodized), while 90° peel tests were carried out to complement the mechanical evaluation. Numerical models in ABAQUS®FE were developed to replicate the experimental tests and validate the observed behaviours.

The results show that adhesive type and surface treatment strongly affect joint strength and failure modes. LS-XU joints consistently failed adhesively, with anodization producing the best performance, whereas Teroson MS 939 promoted cohesive failure under sanding and anodization, resulting in higher strength and energy absorption. Peel tests confirmed the superior viscoelastic dissipation of Teroson MS 939, with higher critical energy release rates (~4.4 N/mm) compared to LS-XU (~3.2 N/mm). These findings highlight the importance of selecting appropriate adhesive systems and surface preparation strategies for cast aluminium in automotive structures, where balancing strength, ductility, and durability is critical to meet lightweighting and performance requirements.

Keywords: Adhesive bonding · hot melt adhesive · silane-modified polymer · aluminium die casting · surface preparation

1 Introduction

Die-cast aluminium alloys, particularly Al–Si–Mg systems such as $AlSi_{10}Mg$, are widely employed in the automotive industry due to their low density, high specific strength, good corrosion resistance, and excellent castability for complex geometries [1–3]. Their

L. F. M. da Silva and R. D. Adams (Eds.): AB 2025, PEM, pp. 141–169, 2026.
https://doi.org/10.1007/978-3-032-19022-2_9

use supports industry-wide lightweighting initiatives aimed at reducing fuel consumption and greenhouse gas emissions. However, joining die-cast aluminium components presents significant challenges compared to wrought alloys. The die-casting process often produces a heterogeneous surface layer containing entrapped oxides, silicon-rich eutectic phases, porosity, and residual release agents. These features reduce surface wettability, hinder the formation of durable adhesive bonds, and may promote interfacial defects under service conditions [4].

Surface preparation is therefore a critical step in achieving reliable bonding performance. Mechanical treatments such as abrasion, sanding, or grit blasting are commonly used to increase surface roughness and promote mechanical interlocking. Chemical treatments, including alkaline cleaning, acid etching, or chromate conversion coatings, are effective in removing contaminants and modifying surface chemistry. More advanced approaches, such as anodizing and plasma treatment aim to increase surface energy and introduce chemically active sites, thereby improving adhesive–substrate interaction and long-term durability [5, 6]. The selection of an appropriate treatment must balance performance benefits with manufacturing feasibility, environmental regulations, and compatibility with downstream processes.

The application of structural adhesive bonding to die-cast aluminium in automotive body-in-white (BIW) assembly remains a relatively recent development compared to its extensive use with sheet-formed aluminium and steel panels. Historically, the complex surface morphology, high silicon content, and susceptibility to casting defects made diecast aluminium less attractive for adhesive bonding, leading manufacturers to rely on mechanical fastening methods such as riveting, bolting, and self-piercing rivets, which add weight, manufacturing cost and constrain design flexibility [3, 7]. The growing adoption of high-performance surface treatments and tailored adhesive formulations has now made it feasible to achieve durable, high-strength bonds in die-cast components, enabling engineers to replace heavy or intrusive fasteners with bonded joints. This shift represents a significant innovation pathway in automotive lightweighting, offering manufacturing advantages, including reduced part count and simplified assembly processes [8].

In this study, the bonding performance of die-cast aluminium $AlSi_{10}Mg$ was evaluated using two adhesive systems: a silane-modified polymer (SMP) sealant and a thermoplastic hot melt adhesive (HMA) cycloolefin-based film. SMPs, also referred to as silyl-modified or MS polymers, are hybrid adhesives combining elasticity, durability, and environmental resistance. According to Henkel [9], SMPs undergo one-component, moisture-curing that yields elastomeric bonds free of solvents, isocyanates, silicones, and PVC, with excellent UV and weather resistance and strong adhesion to diverse substrates, even without primers. This sealant imparts high initial adhesion, excellent sag resistance, and long-term flexibility, characteristics that make SMPs particularly suited to automotive applications requiring durable elastic sealing, vibration damping, and structural bonding of body components [10].

HMA films, on the other hand, soften when heated and solidify on cooling, enabling rapid assembly, consistent bondline thickness, and minimal curing time [11]. They provide high initial tack, good gap filling, and reliable adhesion to various substrates, which suits high-volume automotive manufacturing where process speed and reproducibility are critical [12]. The comparative assessment of these two adhesive systems on $AlSi_{10}Mg$ substrates provides insight into the influence of adhesive type on joint performance and the suitability of each for structural and sealing applications in the automotive sector.

In the automotive sector, adhesive bonding of die-cast aluminium is gaining strategic importance as manufacturers transition toward multi-material body architectures and reduced mechanical fastening. Furthermore, the rise of the megacasting type of structures for body construction reinforces the move away from conventional sheet metal and its associated conventional joining methods, such as bolting, riveting, clinching [13]. In these cases, this imposes strict design constraints, including the need for overlapping flanges, local stress concentrations, added mass, and reduced fatigue life. Adhesive bonding offers unique advantages: it enables uniform load transfer over a large area, seals joints against corrosion, eliminates stress risers, and allows for joining thin or dissimilar materials without introducing heat distortion. These attributes are essential for modern vehicle structures, where weight reduction, crash performance, and production efficiency must be achieved simultaneously. Effective bonding of die-cast aluminium is thus not only a materials engineering challenge but also a key enabler for the next generation of lightweight and sustainable automotive designs [3, 4].

As shown, there is limited information on how to join cast aluminium using adhesive bonding, although the importance of this technique is growing. In this work, experimental and numerical steps were carried out to identify suitable procedures for bonding these materials. In more detail, single lap joints (SLJs) employing die-cast aluminium $AlSi_{10}Mg$ adherends were manufactured using two distinct adhesives: Zeon's hot melt adhesive LS-XU and Henkel's MS polymer Teroson MS 939 adhesive. The adherends underwent three different surface preparation methods-degreasing, sanding, and anodizing—to evaluate their influence on joint performance, with particular attention to failure modes and adhesion characteristics. Additionally, the aluminium substrates were tested in two surface conditions, as-cast and machined, to assess the impact of machining operations on surface properties and adhesion. 90° peel tests were conducted to determine the peel strength of both adhesives under the aforementioned surface preparation and substrate condition scenarios. Complementary numerical models were developed in ABAQUS®FE using geometries replicating those of the experimental specimens, thereby enabling direct correlation between the experimental and numerical results.

2 Experimental details

2.1 Materials

A multi-material film adhesive was used, with the commercial reference LS-XU, with 0.4 mm thickness and cyclic olefin-based polymer produced by Zeon® Corporation (Chiyodaku, Tokyo, Japan). It was cured by using a hot press, under 159 °C during 45 minutes, proceeded by cooling until room temperature. Rodrigues et al. [14] characterised this adhesive at a thickness of 0.1 mm, and a subsequent study on the 0.4 mm variant provided the mechanical properties listed in Table 1.

TEROSON® MS 939 is an adhesive/sealant based on modified silane polymers, provided by Henkel (Du¨sseldorf, Germany). It was cured by absorbing ambient humidity,

Table 1. LS-XU mechanical properties

Property	Units	Value
Young's modulus, E	[MPa]	230.18 ± 8.08
Yield strength, σy	[MPa]	5.49 ± 0.36
Tensile failure strength, σf	[MPa]	10.86 ± 1.48
Tensile failure strain, εf	[%]	1066 ± 83
Poisson's ratio, v	[-]	0.47 ± 0.02
Shear modulus, G	[MPa]	76.67 ± 2.52
Shear failure strength, τf	[MPa]	5.09 ± 0.08
Shear failure strain, γf	[%]	354 ± 19
Toughness in mode I, JIc	[N/mm]	0.20 ± 0.05
Toughness in mode II, $JIIc$	[N/mm]	2.66 ± 0.19

during two weeks, forming an elastic product. Table 2 presents the mechanical properties of this MS polymer adhesive.

Table 2. Teroson MS 939 mechanical properties

Property	Units	Value
Young's modulus, E	[MPa]	2.90 ± 0.01
Yield strength, σy	[MPa]	8.24
Tensile failure strength, σf	[MPa]	3.14 ± 0.03
Tensile failure strain, εf	[%]	2.30 ± 0.04
Poisson's ratio, v	[-]	0.37
Shear modulus, G	[MPa]	0.94 ± 0.02
Shear failure strength, τf	[MPa]	2.59 ± 0.07
Shear failure strain, γf	[%]	5.68 ± 0.36
Toughness in mode I, JIc	[N/mm]	5.2 ± 1.8

The substrates are made of the aluminium alloy $AlSi_{10}Mg$, extracted from a high pressure die-cast automotive part, and, therefore in the cast state. The Young's modulus, E, considered is of 70 GPa and a Poisson's ratio, v, of 0.33.

2.2 Specimen Manufacturing

Four types of single lap joints were manufactured for the purpose of studying their joint behaviour, with two different adhesives and two different substrates. The substrates of $AlSi_{10}Mg$ were machined to obtain the dimensions 10 x 100 x 3 mm, and were considered in the as-cast and machined state. All joints have an overlap of 25 mm and 0.2 mm of adhesive thickness. Figure 1 presents the dimensions of the single lap joints.

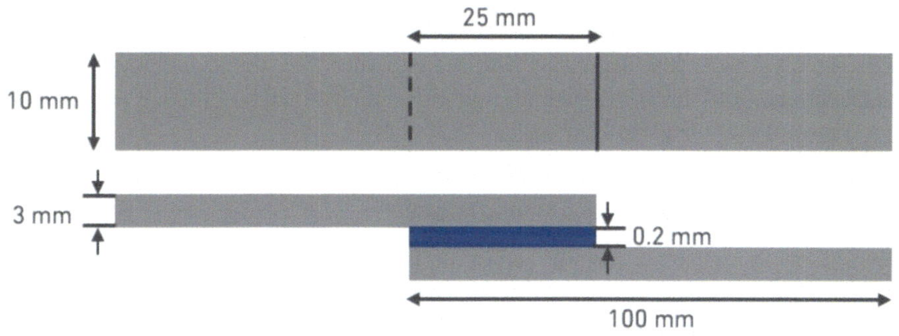

Fig. 1. Single lap joint geometry used.

2.3 Surface Energy and Surface Preparation

As a preliminary test, the surface energy was measured to understand how properties like adhesion, wettability, and coating compatibility behave in these substrates, as cast and machined. The test consists in the direct measurement of the contact angle of different fluids using a OCA 15 goniometer (DataPhysics, Neurtek Instruments, Eibar, Spain) under ambient conditions, using three different liquids: water (polar liquid), ethylene glycol 55% (polar liquid) and n-hexadecane (nonpolar). The surface energy is obtained by combining the Owens-Wendt-Rabel-Kaelble (OWRK) method and the measured contact angles. In Figure 2 are presented the results of this measurement.

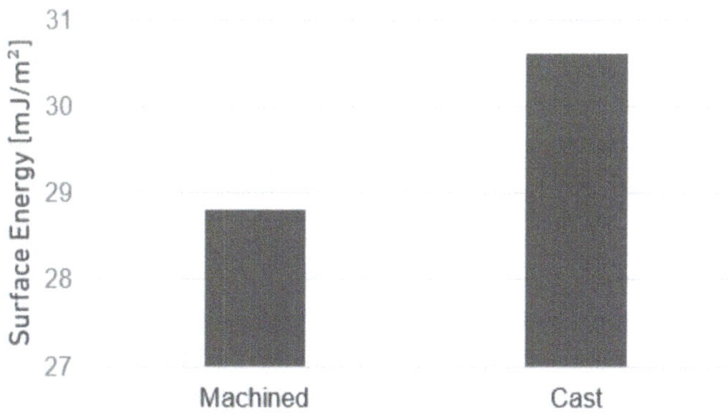

Fig. 2. Surface energy of substrates in the as-cast and machined states [3].

The theoretical surface energy of aluminium lies around 500 to 1000 mJ/m^2, whereas real (passivated) surfaces range from 40 to 60 mJ/m^2, depending on the surface roughness, cleaning procedure and alloy constituents [15]. The results presented are lower than the ones found in the literature, which can be attributed to the fact that the material in study is die-cast aluminium, rather than pure aluminium with implications on the surface topography and the eventual presence of contaminants which are not fully removed by an acetone pre-cleaning [3].

Regarding the surface preparation, three different treatment surfaces were considered. The substrates were degreased with acetone, then they were sanded with a p600 sandpaper and lastly anodized with phosphoric acid anodization. Figure 3 presents the appearance of the substrates after surface treatments.

(a) Degreased (b) Sanded (c) Anodized

Fig. 3. Substrates appearance after the surface treatment.

2.3.1 As Cast Aluminium Anodization

Anodizing as-cast aluminium presents significant challenges due to the surface characteristics and alloy composition of the material. In contrast to wrought aluminium, as-cast alloys frequently exhibit elevated levels of silicon and other alloying elements, resulting in a rough, porous, and heterogeneous surface. The irregular surface of the material in question has been shown to interfere with the formation of a uniform and adherent anodic oxide layer. This has been shown to result in poor corrosion resistance, inconsistent colouring, and reduced aesthetic quality. Intermetallic particles and casting defects disrupt the anodizing process, making it difficult to achieve reliable, high-quality finishes on aluminium parts [16]. In Fig. 4 are presented different results of anodized as-cast aluminium substrates, that were anodized under the same conditions, at the same time.

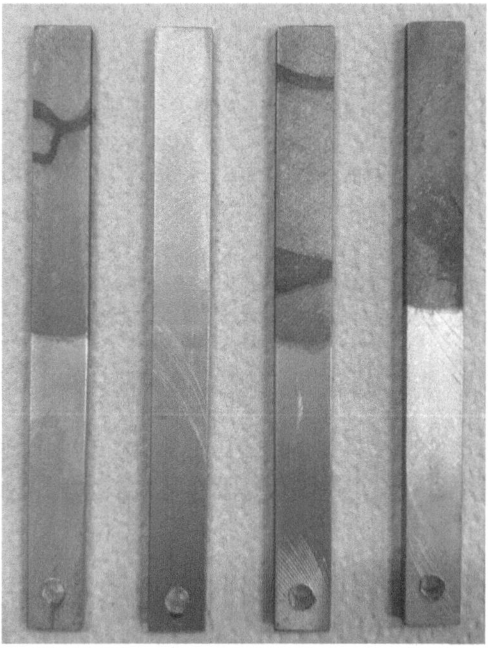

Fig. 4. Different results of the anodization of as-cast aluminium substrates [3].

2.4 Testing Procedures

Single Lap Joints

The single lap joints were tested at the rate of 1 mm/min (quasi-static), to avoid inertial and dynamic effects, like acceleration, vibrations, or strain-rate sensitivity. The tests were performed on the testing machine INSTRON® 3367 universal (Illinois ToolWorks, Hopkinton, MA, USA).

90° Peel Test

The 90° peel test was performed as described in ASTM D6862-11 [17], in the testing machine INSTRON® 3367 universal (Illinois ToolWorks, Hopkinton, MA, USA) and it is a method used to evaluate the adhesive strength between two bonded materials, in this case the aluminium alloy $AlSi_{10}Mg$ and both adhesives. In this test, one adherent, the peel arm, which is considered the flexible adherend, was clamped vertically in the upper grip. The other adherent, which is considered the rigid adherend, was fixed horizontally in the lower grip to maintain a 90° peel angle during testing. The peel arm is pulled at a right angle (90°) at 12 mm/min to the bonded interface while the other remains fixed, allowing the measurement of the force required to progressively separate the layers. This test provides insight into the adhesion performance and failure characteristics of the adhesive.

For this test, the only surface treatment applied to the adherends was acetone degreasing.

3 Numerical Details

To simulate the behaviour of the joints with the LS-XU film adhesive and the Teroson MS 939 MS polymer, the software ABAQUS®FE was used to model single lap joints. For both models, it was developed a 2D static model, that is shown in Figure 5. Under quasistatic conditions, the single lap joints testing procedures were modelled using ABAQUS® CPS4R plane stress elements.

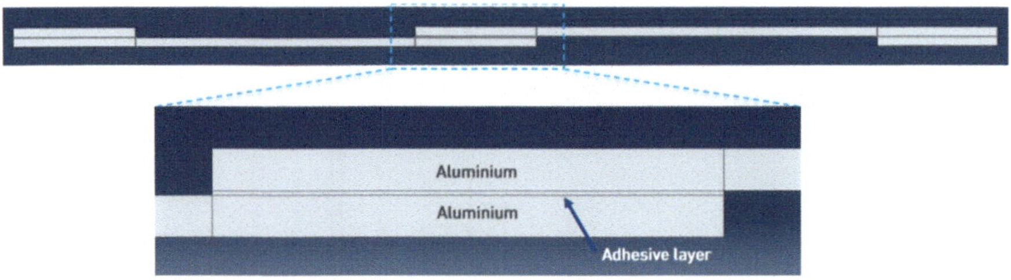

Fig. 5. 2D static model of the SLJ in ABAQUS®FE.

The boundary conditions were established according to the experimental tests of single lap joints. The left end of the joint was pinned (U1=U2=U3=0) and a displacement (U1) was applied to the right end of the joint, as shown in Figure 6.

Fig. 6. Boundary conditions applied to the SLJ in ABAQUS®FE.

Regarding the mesh applied to the model, it is refined in the regions with higher stress concentrations, in the proximity of the overlap edges. The overlap area is characterised by the presence of two distinct element sizes: smaller elements positioned at the periphery and larger elements situated within the central region. Then, until both end sides of the SLJ, the mesh increases again. This approach ensures computational efficiency by avoiding over-meshing the part. As illustrated in Figure 7, mesh refinement is evident in the overlap regions.

Fig. 7. Mesh applied to the SLJ in ABAQUS®FE.

3.1 MS Polymer Adhesive Model

When modelling the Teroson MS 939 MS polymer adhesive, the adhesive layer was characterized with hyperelastic properties. To model the ideal single lap joint failure, which is cohesive, a cohesive layer was introduced between the hyperelastic bulk material, that uses a 4-node two-dimensional cohesive element, COH2D4. Figure 8 presents the schematic representation of the Teroson MS 939 single lap joint in ABAQUS®.

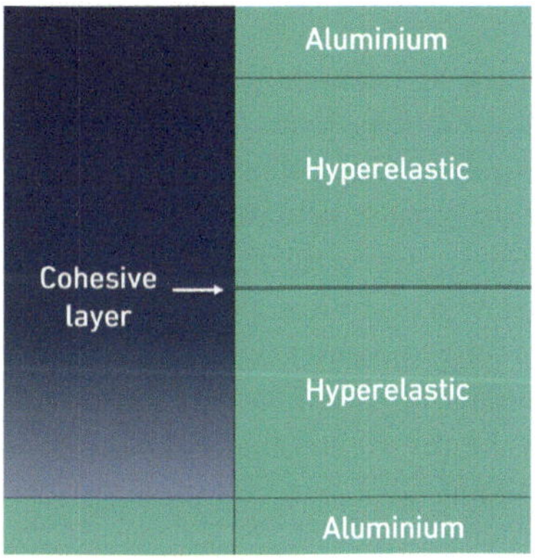

Fig. 8. Schematic representation of Teroson MS 939 single lap joint model as constructed in ABAQUS®.

The hyperelastic model used was Marlow Model, where only one set of test data is provided (e.g uniaxial tension), and the model follows the provided data exactly. For this case, the stress vs strain curve obtained from bulk testing was used as the provided test data. The stress vs strain curve is presented in Figure 9. The cohesive layer has the elastic properties of the adhesive, and the damage initiation followed the quadratic nominal stress criterion (QUADS) in ABAQUS®.

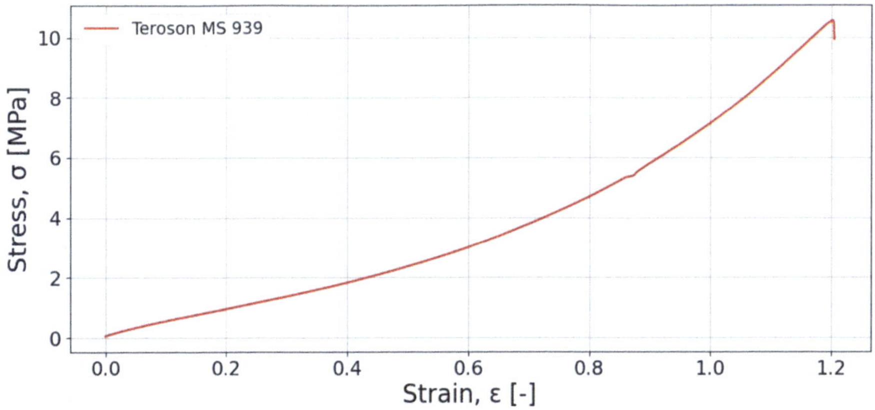

Fig. 9. Stress vs strain curve for Teroson MS 939 obtained from bulk testing.

3.2 Film Adhesive Model

Regarding the modelling of the film adhesive LS-XU, the the bulk adhesive layer was characterized with elastic and plastic properties. The elastic data was obtained from bulk tensile testing and the plastic data from a reverse engineering process, obtaining the stress vs strain curve.

To model the failure of the SLJ, which is adhesive, the debonding is captured with surface-to-surface contact interactions, in each interface adherend-adhesive in the joint, provided with cohesive behaviour. The cohesive properties were inputted as a contact property defined with traction-separation elastic stiffness and damage initiation following the quadratic nominal stress criterion (QUADS) in ABAQUS®. For damage evolution Benzeggagh–Kenane (BK) law was chosen due to the fact that its coefficient, η, controls how the effective fracture energy increases with shear content. This matches the experimentally observed rise in toughness for hot-melt adhesives under mode-II-dominated loading in single-lap joints. Figure 10 presents the schematic representation of the film adhesive LS-XU single lap joint.

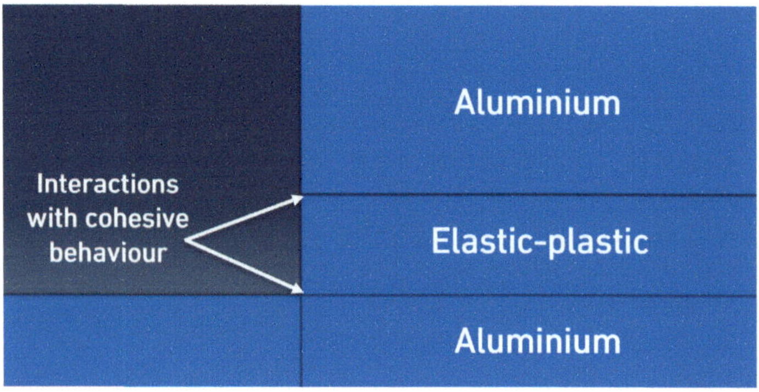

Fig. 10. Schematic representation of LS-XU single lap joint in ABAQUS®.

4 Results

4.1 Single Lap Joints

For the SLJs testing, at least three specimens were considered in the presented results. Figure 11 shows the results of the tested SLJs with the adherends in as-cast condition, treated with the three different surface treatments previously mentioned, and the adhesive film LS-XU. These results highlight the significant influence of surface preparation on joint strength, with anodizing providing the most favourable mechanical performance, reaching the highest ultimate load of around 1400 N at ~ 1.35 mm of displacement. Then, it is followed by sanding, that achieves around 950 N of highest load at ~ 1.05 mm of displacement. Finally, degreasing produced the weakest joints, only reaching 900 N at ~ 0.80 mm of displacement.

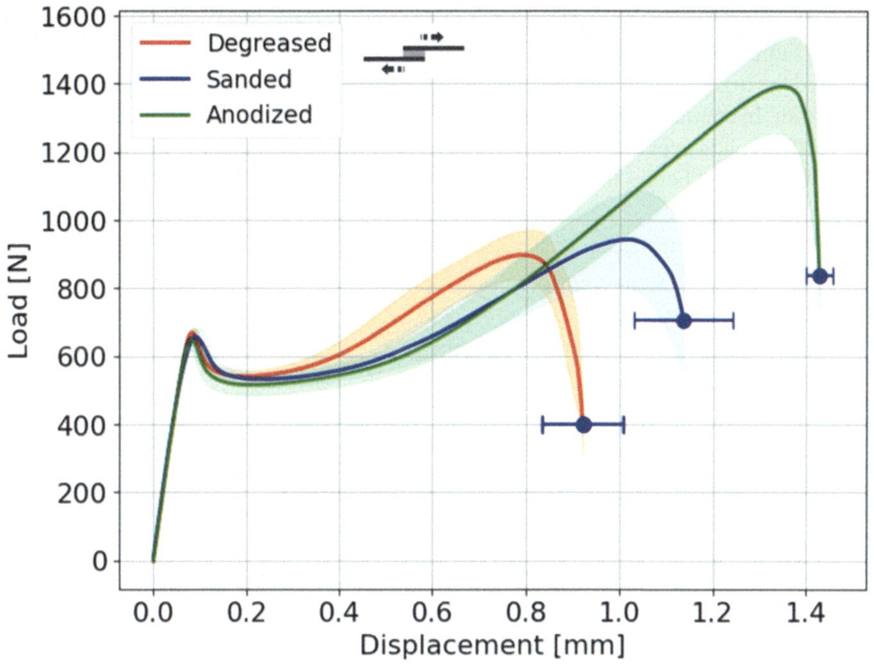

Fig. 11. Results of SLJs with substrates in as-cast condition and the LS-XU film adhesive [3].

Figure 12 shows that all tested SLJs failed by adhesive failure, meaning the fracture occurred at the interface between the adhesive and the substrate. In this failure mode, the adhesive cleanly detaches from one adherend, leaving little or no residue, indicating that interfacial adhesion was the limiting factor, rather than the adhesive's internal strength or the substrate's strength.

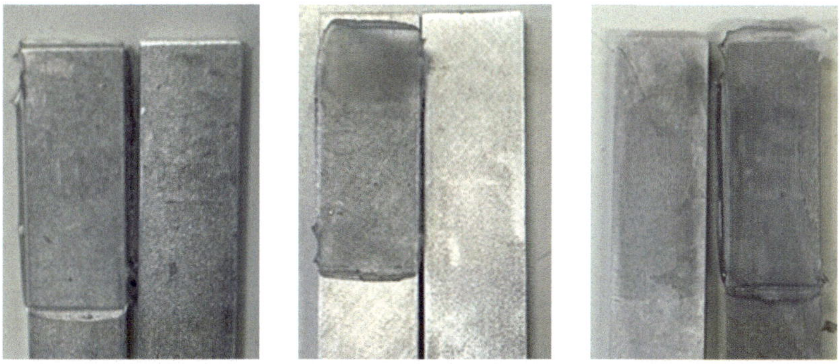

(a) Degreased (b) Sanded (c) Anodized

Fig. 12. Failure surfaces of the SLJ's with as-cast substrate and the film adhesive [3].

For the substrates in the machined condition, Fig. 13 shows the results for SLJs tested with the same film adhesive LS-XU, and the same surface treatments. The anodized SLJs exhibited the highest ultimate load, attaining approximately 1450 N at a displacement of ~ 1.10 mm. In contrast, both sanded and degreased SLJs failed at a similar displacement of ~ 0.80 mm, however, the sanded specimens achieved a substantially higher peak load of about 1200 N, whereas the degreased specimens reached only around 950 N.

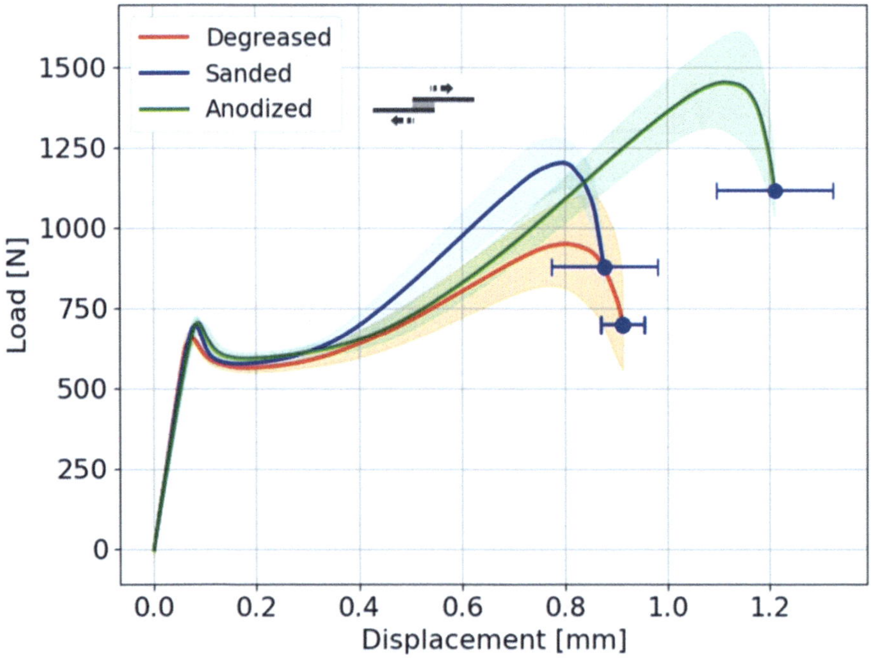

Fig. 13. Results of SLJs with substrates in machined condition and the LS-XU film adhesive [3].

Figure 14 shows the type of failure that occurred in the joint, it was adhesive failure in all of them, as for the as-cast condition previously presented.

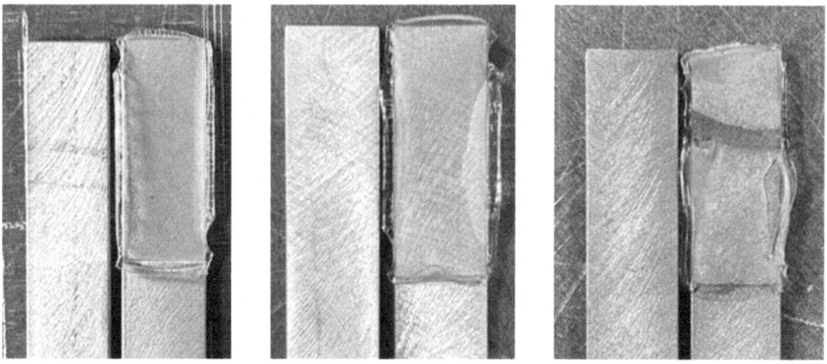

(a) Degreased (b) Sanded (c) Anodized

Fig. 14. Failure surfaces of the SLJs with machined substrates and the LS-XU film adhesive [3].

After testing both specimens conditions, as-cast and machined, with three surface treatments, it is possible to compare both conditions for the same surface treatment. This comparison between as-cast and machined substrates highlights the influence of surface condition on the mechanical performance and failure behaviour of SLJs.

Figure 15 presents the comparison between substrates' surface conditions as-cast and machined when degreased with acetone, using the LS-XU film adhesive. Both substrates' conditions exhibit very similar behaviour, with peak loads approaching 900–950 N and failure occurring at comparable displacements ~ 0.80 mm. The close overlap between the curves suggests that machining offers no significant benefit when the failure mechanism is dominated by interfacial debonding. The reduced effectiveness of surface preparation in this case can be explained by the limiting factor that is the adhesive–substrate interface rather than the bulk adhesive properties.

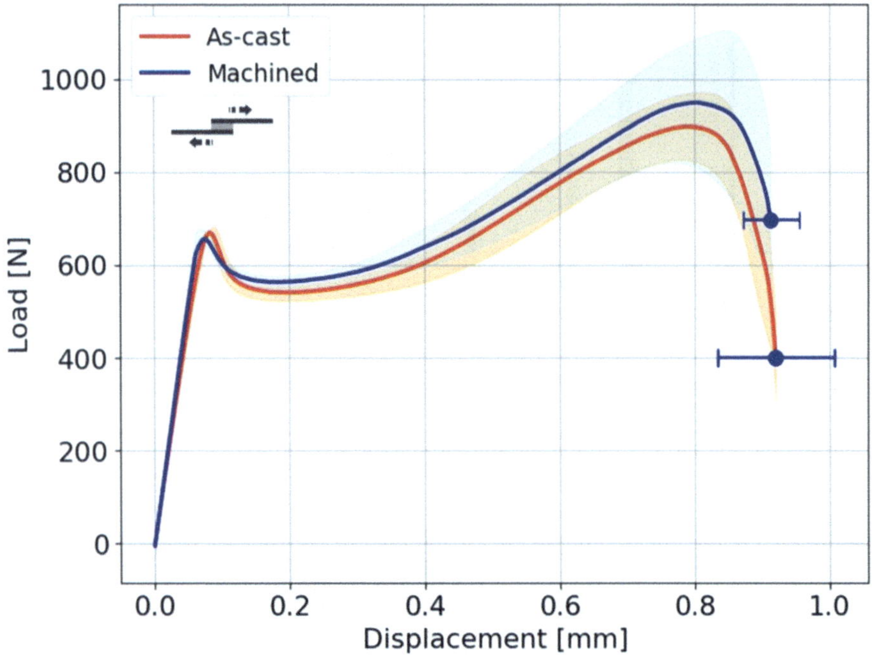

Fig. 15. Comparison between substrates' surface condition when degreased, using the LS-XU film adhesive [3].

Regarding sanding as surface treatment, Figure 16 presents the comparison between as-cast and machined conditions for the adhesive film LS-XU. For this surface treatment there is a stronger difference in the results, whereas the machined substrates achieve a much higher peak load around 1200 N, at a lower displacement of ~ 0.80 mm, than the as-cast substrates that only reach around 950 N, at ~ 1.05 mm of displacement. This improvement can be attributed to the combined effect of sanding and machining, which reduces surface irregularities while increasing effective contact area, thereby promoting stronger interfacial adhesion. In contrast, the as-cast specimens, despite the sanding treatment, may be affected by residual surface defects that continue to act as stress concentrators.

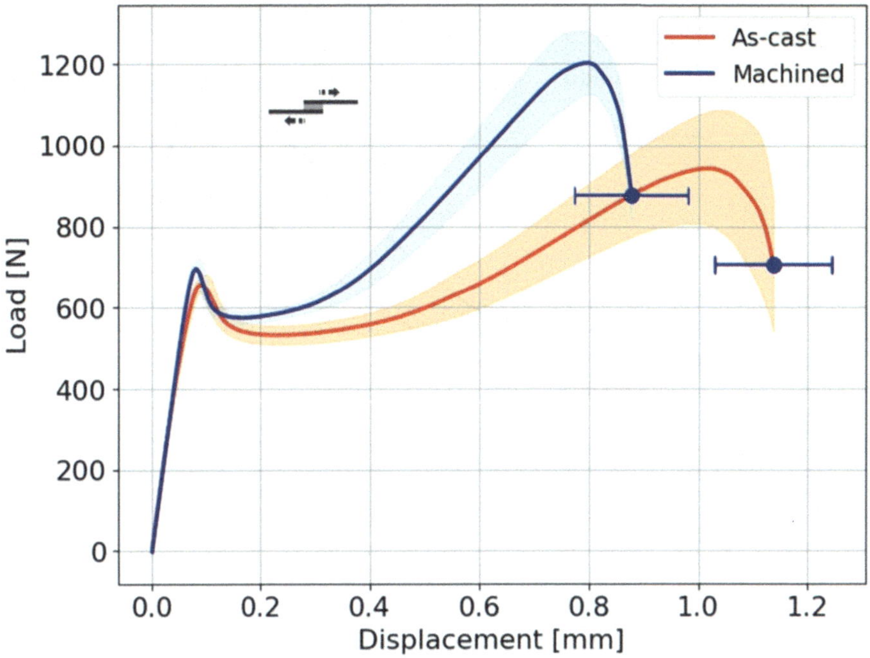

Fig. 16. Comparison between substrates' surface condition when sanded, using the LSXU film adhesive [3].

Finally, for the anodized specimens, Figure 17 presents the comparison between as-cast and machined conditions for the adhesive film LS-XU. The machined specimens achieve the highest peak strength, exceeding 1400 N, whereas the as-cast specimens reach slightly lower values. However, the as-cast anodized joints sustain greater displacements before failure, extending beyond 1.4 mm compared to approximately 1.1 mm for the machined condition. This suggests a trade-off between strength and ductility: machining prior to anodization enhances maximum load capacity, likely due to improved surface uniformity and oxide layer quality, but the as-cast surface retains features that promote slightly greater deformation before interfacial debonding. Overall, anodization provides the most effective treatment, with the as-cast and machined variants favouring ductility and strength, respectively.

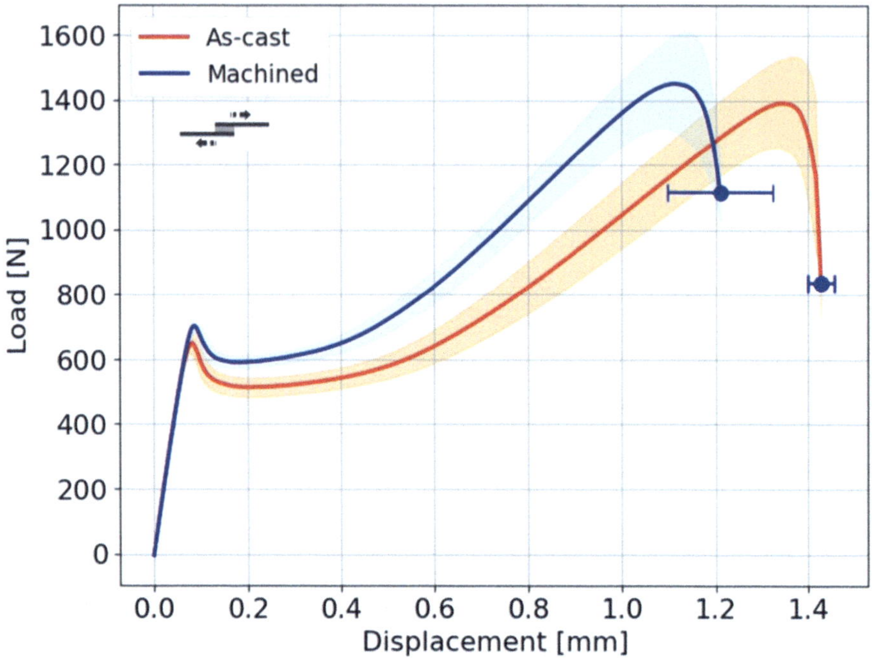

Fig. 17. Comparison between substrates' surface condition when anodized, using the LS-XU film adhesive [3].

Figure 18 shows the results of the tested single lap joints with substrates in as-cast condition and with MS polymer adhesive Teroson MS 939. It can be seen that the surface treatment clearly influences the maximum load reached in the SLJs, being the anodized specimens the ones with better outcome.

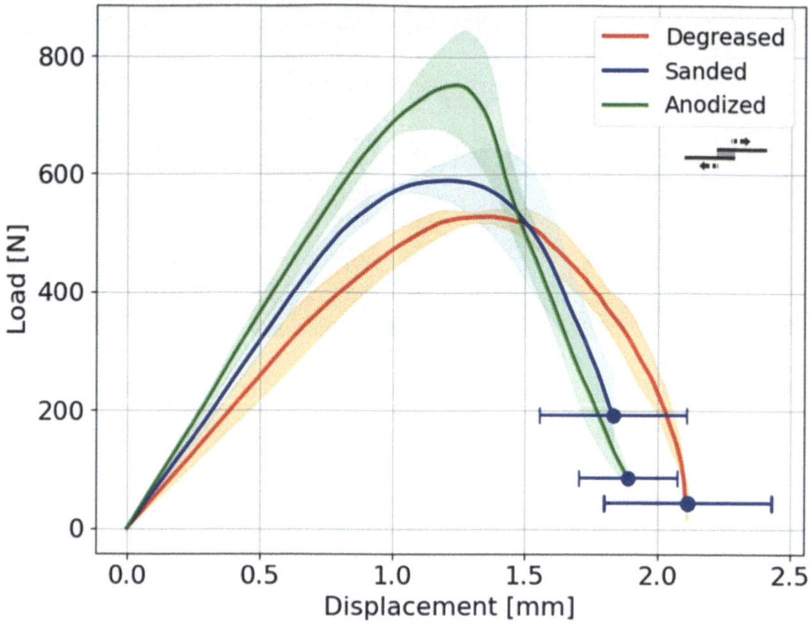

Fig. 18. Results of SLJs with substrates in as-cast condition and the Teroson MS 939 adhesive [3].

Figure 19 illustrates the failure mechanisms observed for each surface treatment. In the case of the degreased specimens, a mixed mode failure was identified, with adhesive failure being predominant. This behaviour accounts for the comparatively lower load-bearing capacity of the degreased SLJs. For the sanded specimens, mixed failure was also observed, however, cohesive failure was more prevalent. This results in a higher peak load than the degreased specimens. The anodized specimens predominantly exhibited cohesive failure, which represents the most favourable outcome for MS polymer SLJs and is consistent with their superior maximum load capacity.

Fig. 19. Failure surfaces of the SLJs with as-cast substrates and the Teroson MS 939 adhesive [3].

Figure 20 shows the results of the tested single lap joints with substrates in machined condition and with the Teroson MS 939 adhesive. As observed in the results for the as-cast condition, the surface treatment significantly affects the maximum load attained by the SLJs. Among the treatments, anodized specimens demonstrate the highest performance, although sanded specimens achieve comparably close values.

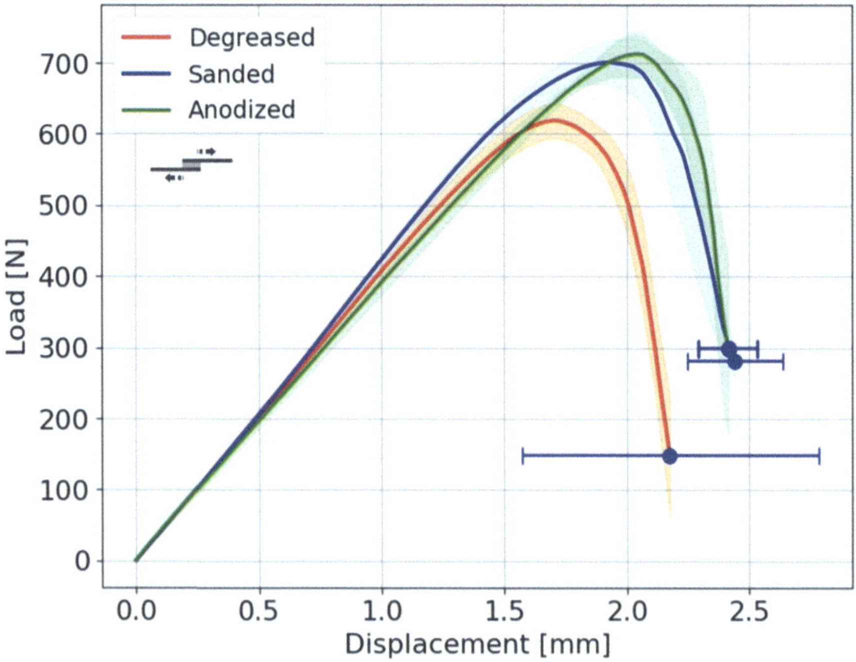

Fig. 20. Results of SLJs with substrates in machined condition and the Teroson MS 939 adhesive [3].

Figure 21 shows the type of failure that occurred in the joint, for each surface treatment. It can be concluded that the failure modes observed in the machined substrates are consistent with those identified in the as-cast substrates. The sanded specimens exhibit an improved failure behaviour, characterised by a greater proportion of cohesive failure, which accounts for their performance approaching that of the anodized specimens.

(a) Degreased (b) Sanded (c) Anodized

Fig. 21. Failure surfaces of the SLJs with machined substrate and the Teroson MS 939 adhesive [3].

Again, after testing both specimens conditions, as-cast and machined, with three surface treatments, it is possible to compare both conditions for the same surface treatment, this time for the MS polymer adhesive.

Figure 22 presents the comparison between as-cast and machined conditions for degreased adherends. The machined specimens achieve a higher maximum load, exceeding 600 N, and sustain larger displacements prior to failure, indicating improved ductility and energy absorption. This behaviour can be explained by a higher predominance of cohesive failure in machined specimens, which is generally associated with stronger interfacial adhesion. In contrast, the as-cast specimens peak at around 520 N and fail at lower displacements, consistent with a greater proportion of adhesive failure. This reduction in performance can be attributed to surface irregularities and defects inherent to the as-cast condition, which promote stress concentrations and weaken the joint interface.

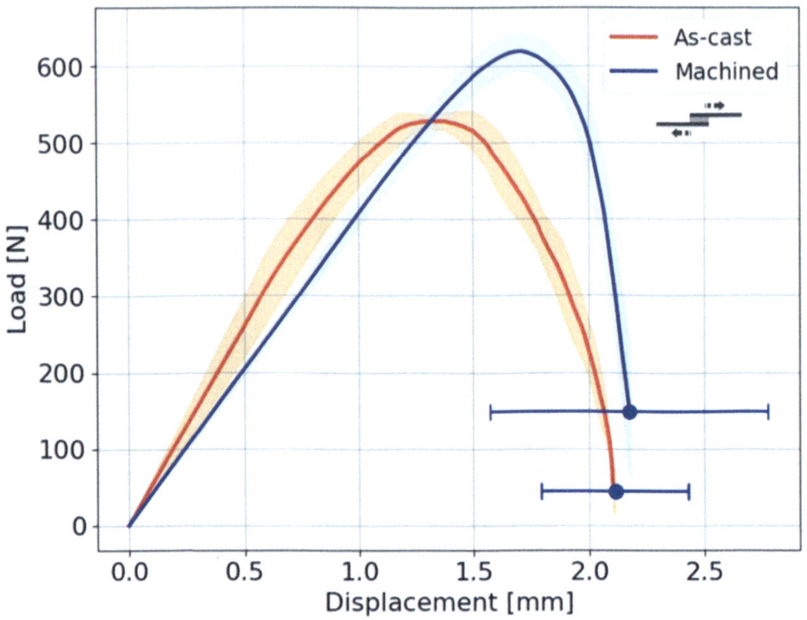

Fig. 22. Comparison between substrates' surface condition when degreased, using the Teroson MS 939 adhesive [3].

Figure 23 presents the comparison between as-cast and machined conditions for sanded adherends. This comparison demonstrates a marked improvement in joint performance for the machined condition. The machined specimens achieve a significantly higher maximum load, exceeding 700 N, compared to approximately 600 N for the as-cast counterparts, as of a larger displacement prior to failure. These results are consistent with the predominance of cohesive failure observed in sanded specimens, which promotes stronger interfacial bonding. In contrast, the as-cast specimens, although improved compared to untreated surfaces, still display earlier failure associated with residual surface irregularities and localised stress concentrations. Overall, sanding combined with machining leads to superior mechanical behaviour, with performance.

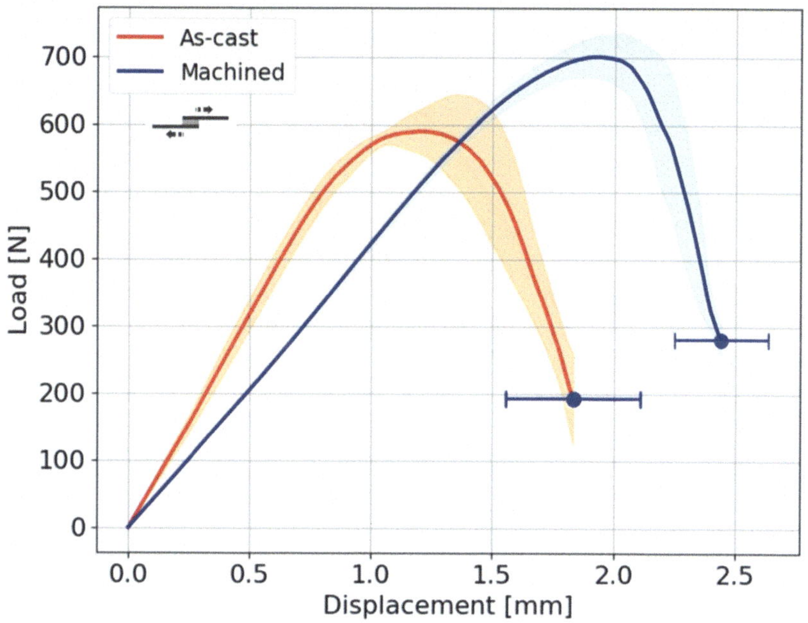

Fig. 23. Comparison between substrates' surface condition when sanded, using the Teroson MS 939 adhesive [3].

Figure 24 presents the comparison between as-cast and machined conditions for anodized adherends. Despite their lower peak load, the machined joints sustain larger displacements before failure, indicating superior ductility and energy absorption. This divergence suggests that while anodization enhances adhesion strength in both conditions, the as-cast surface morphology may provide additional mechanical interlocking that favours higher strength, whereas machining prior to anodization promotes more stable cohesive failure and delayed failure onset. The predominance of cohesive failure in both cases confirms anodization as the most effective surface treatment, though the balance between strength and ductility varies depending on whether machining is applied beforehand.

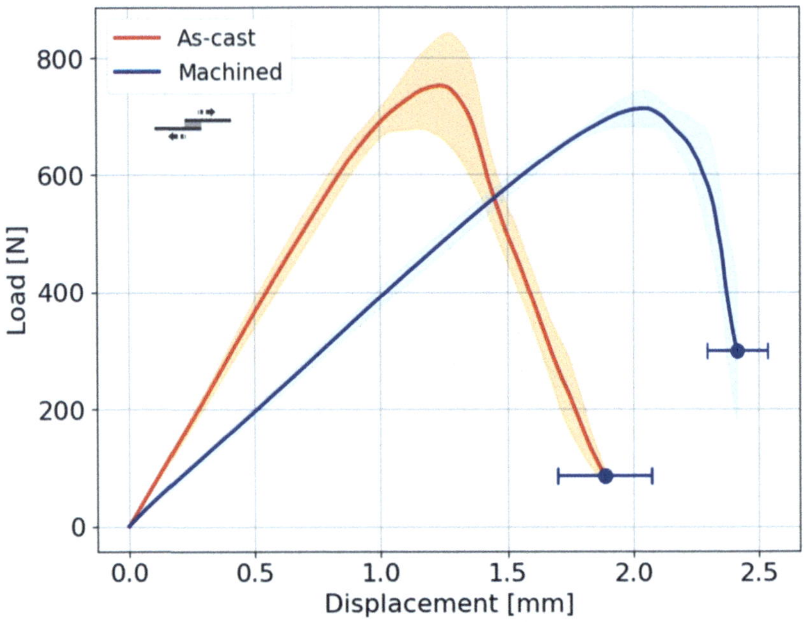

Fig. 24. Comparison between substrates' surface condition when anodized, using the Teroson MS 939 adhesive [3].

4.2 90° Peel Test

As stated before, the 90° peel test involves peeling a flexible steel strip bonded with adhesive to a rigid substrate at a 90° angle. Figure 25 shows the load vs displacement curve obtained directly from the test machine, for the LS-XU film adhesive. For this result, three tests were performed. The peel strength of this film adhesive is of $\sim$ 23 N.

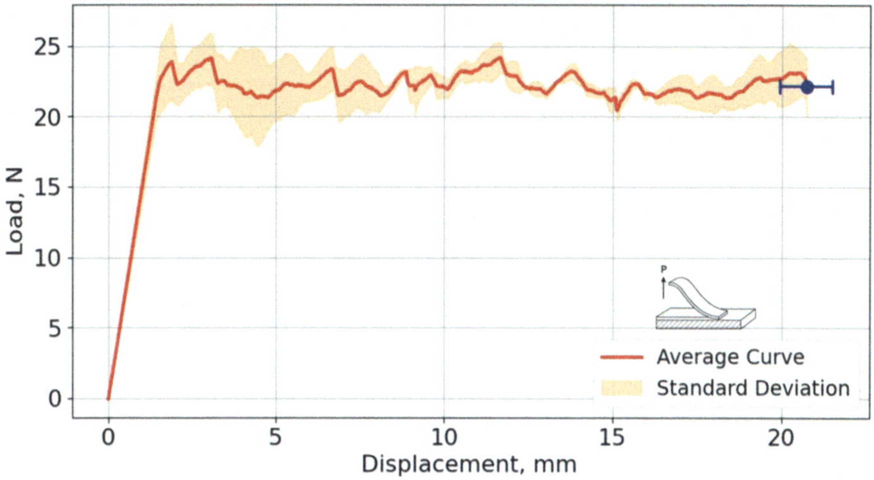

Fig. 25. Results of the peel test of the LS-XU film adhesive [3].

The critical energy release rate, G (in N/mm), is determined from the load vs displacement data using the relation $G = \frac{P}{b}$, where P is the peel force and b is the bond width. By plotting G against the corresponding displacement, a curve is obtained that characterizes the fracture behaviour of the adhesive. This method ensures that any plastic deformation of the substrate is accounted for, allowing G to reflect the true energy required for adhesive fracture [18]. Figure 26 shows the curve that describes the fracture behaviour of the LS-XU film adhesive. The critical energy release rate, G, for the adhesive film is of $\sim$3.2 N/mm.

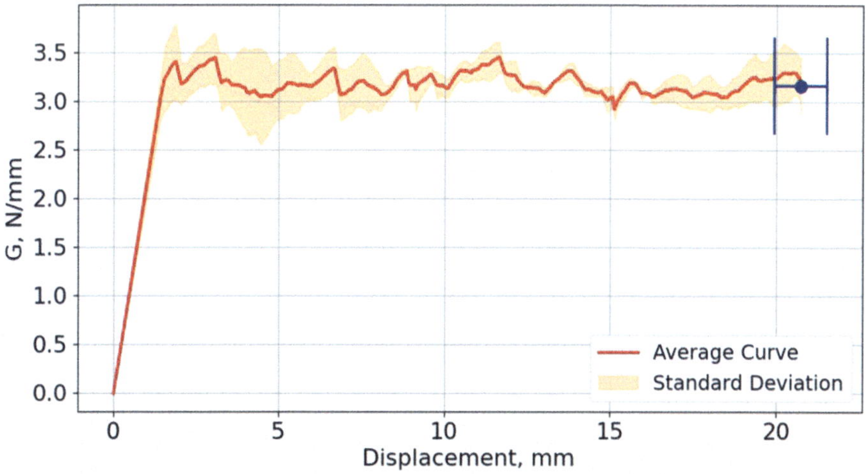

Fig. 26. Critical energy release rate, G, for the LS-XU film [3].

The same test was performed for the MS polymer adhesive, Teroson MS 939. Figure 27 shows the load vs displacement curve obtained directly from the test machine. Three tests were considered to obtain the curve. The peel strength of this MS polymer adhesive is of ~31.6 N.

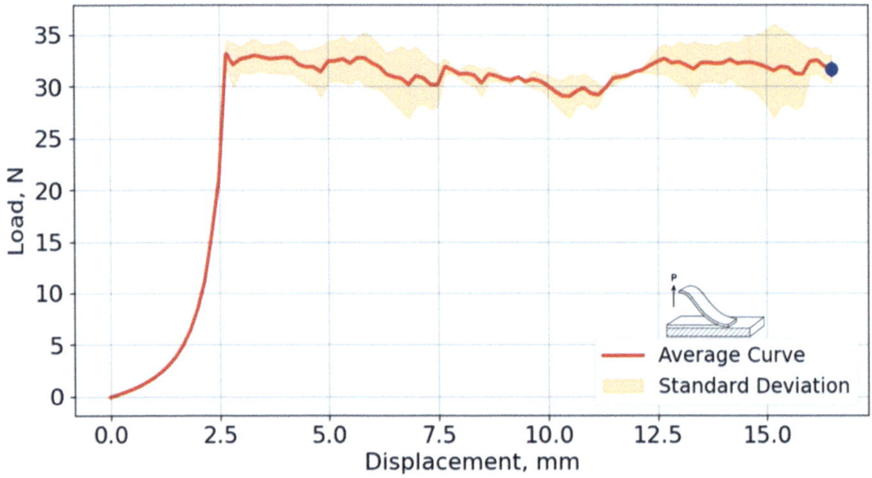

Fig. 27. Results of the peel test of the Teroson MS 939 adhesive [3].

Repetition of the calculations mentioned previously, Figure 28 shows the curve that describes the fracture behavior of the MS polymer adhesive Teroson MS 939. The critical energy release rate, *G*, for the Teroson MS 939 adhesive is of ~ 4.4 N/mm.

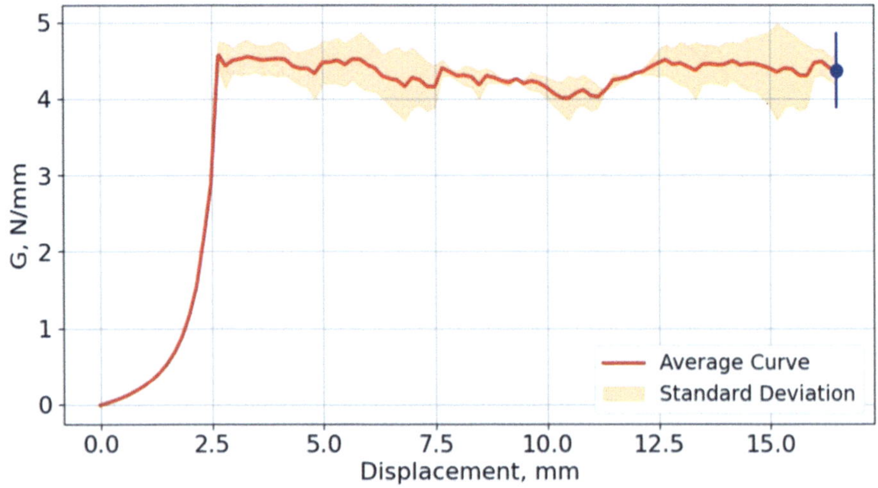

Fig. 28. Critical energy release rate, *G*, for the Teroson MS 939 adhesive [3].

4.3 Numerical Validation

Figure 29 shows the comparison between the experimental curve for the LS-XU adhesive film, considering SLJs with anodized substrates and the numerical model.

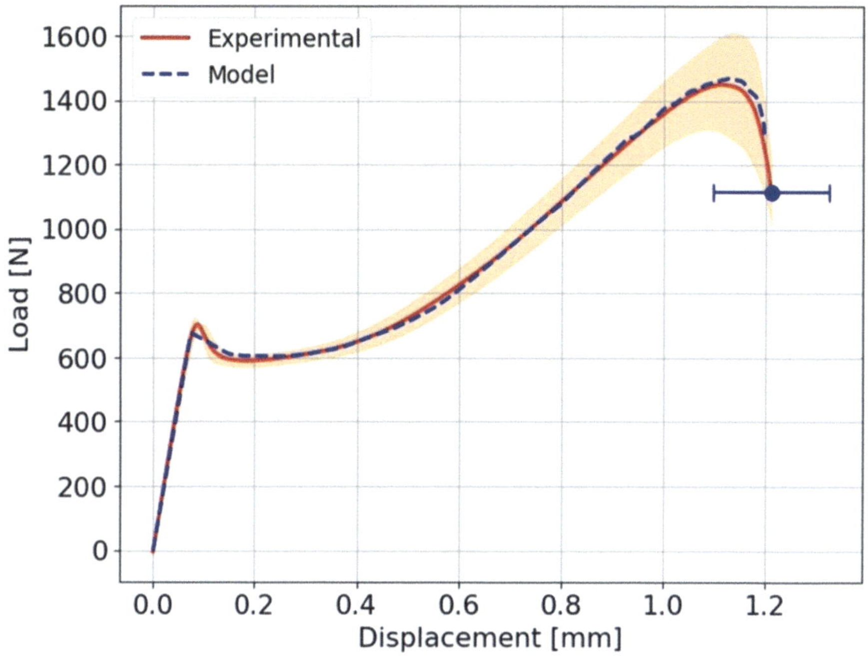

Fig. 29. Comparison between experimental and numerical curves for the LS-XU adhesive film [3].

To obtain these results, an iterative process was taken, considering as reference the stress-strain curve of the adhesive film. This happened because of the need for adjustments in the displacement of the SLJs verified during testing. The adjusted stress-strain curve used in this model is presented in Figure 30.

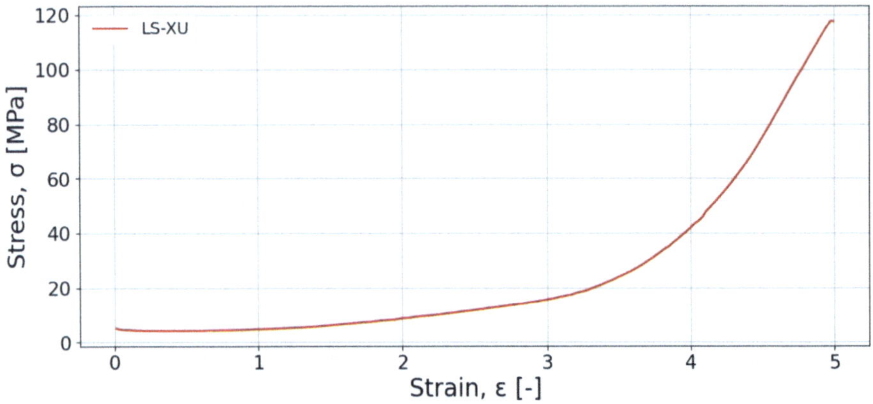

Fig. 30. Adapted stress vs strain curve for the modelling of the LS-XU adhesive [3].

Figure 31 shows the comparison between the experimental curve for SLJs bonded with the Teroson MS 939 adhesive and anodized substrates where the failure was totally cohesive, and the numerical model.

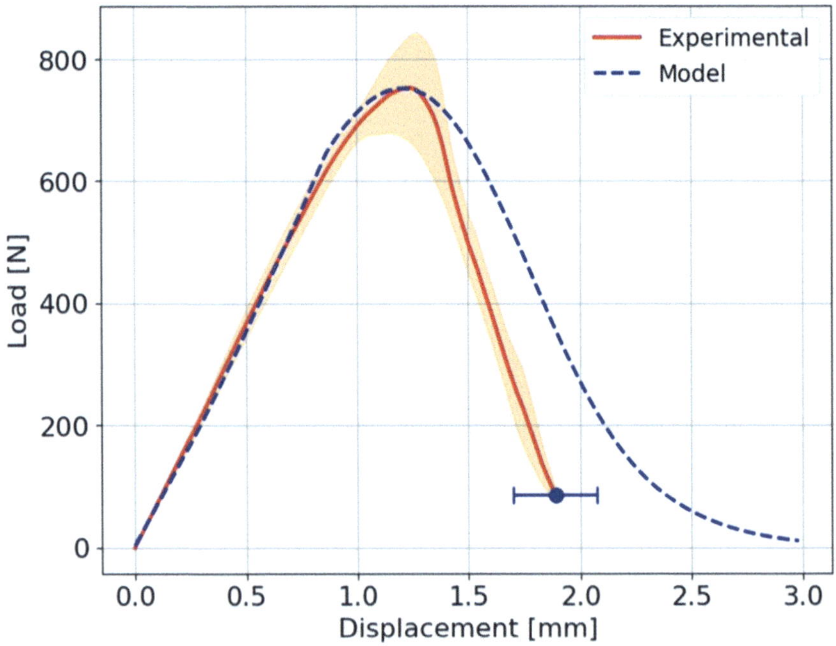

Fig. 31. Comparison between experimental and numerical curves for the Teroson MS 939 adhesive [3].

5 Discussion

The results show that both adhesive type and surface treatment strongly affect SLJ performance. For the MS polymer, degreased substrates exhibited mainly adhesive failure and the lowest strength, sanding promoted more cohesive failure and higher loads, and anodized substrates achieved predominantly cohesive failure with the best performance. This confirms that cohesive failure is the most favourable outcome, as it reflects stronger interfacial bonding and greater energy absorption.

With the adhesive film, failure was exclusively adhesive across all treatments, which limited performance. Nevertheless, surface treatment produced noticeable differences. Degreased specimens showed the weakest behaviour, whereas sanding enhanced both strength and displacement capacity, particularly in the machined state. Anodization produced the best overall performance, with machined substrates reaching the highest loads and as-cast substrates achieving slightly lower strength while sustaining larger displacements.

A clear distinction emerges when comparing as-cast and machined substrates. Machining generally enhanced joint strength by removing surface irregularities and promoting a more uniform stress distribution, an effect most evident in the MS polymer and sanded joints where higher loads were achieved. In anodized adhesive film joints, however, the relationship was more complex: machining maximised strength but reduced displacement, while the as-cast condition supported slightly greater deformation before failure. This indicates that machining improves surface uniformity and oxide quality, whereas the residual morphology of the as-cast surface may provide mechanical interlocking that enhances ductility.

The 90° peel tests indicate that the MS polymer joints absorbed significantly more energy per unit width than the film adhesive joints. The difference arises from the failure mechanisms and adhesive rheology. In the adhesive film, failure was dominated by interfacial debonding, producing relatively constant but lower values of G. This behaviour agrees with early peel models [19, 20], which showed that stiff, low-dissipation adhesives have peel energies governed mainly by the interface. By contrast, MS polymer adhesives are highly viscoelastic and dissipate significant bulk energy during crack advance, raising the steady-state G. This observation is consistent with Creton and Ciccotti [21], who noted that soft adhesives combine interfacial toughness with bulk viscoelastic dissipation.

6 Conclusions

Overall, the findings highlight that the choice between as-cast and machined conditions is not absolute but depends on the desired balance between strength and deformation capacity. For applications where maximum load is critical, machining offers a clear advantage, particularly when combined with anodization. Conversely, where greater ductility and energy absorption are required, as-cast anodized substrates may provide a more favourable outcome.

The 90° peel tests demonstrated that adhesive behaviour under peeling is governed not only by interfacial adhesion but also by the adhesive's rheology. The adhesive film exhibited stable but relatively low peel load and G values, consistent with brittle interfacial debonding, while MS polymer adhesive reached higher G values owing to its bulk viscoelastic dissipation.

This distinction is highly relevant for the automotive industry, where peel resistance is critical in sealing systems, trims, and protective layers exposed to dynamic loads and environmental stresses. MS polymer adhesives, with their higher energy absorption and tolerance to interfacial flaws, provide greater reliability for flexible joints and sealing applications, whereas adhesive films remain advantageous for rigid, dimensionally stable assemblies that demand structural precision rather than peel toughness.

References

1. J. Davis, R.: Aluminum and Aluminum Alloys. ASM International (1993)
2. Kaufman, J.G.: Properties of Aluminum Alloys: Tensile, Creep, and Fatigue Data at High and Low Temperatures. ASM Int. (1999)
3. Ana Almeida Correia. Development and performance evaluation of innovative joining processes for cast aluminium components. Master's thesis, Universidade do Porto (Portugal) (2025)
4. Baser, T.A., Umay, E., Akinci, V.: New trends in aluminum die casting alloys for automotive applications. Eurasia Proc. Sci. Technol. Eng. Math. **21**, 79–87 (2022)
5. Critchlow, G.W., Brewis, D.M.: Review of surface pretreatments for aluminium alloys. Int. J. Adhes. Adhes. **16**(4), 255–275 (1996)
6. Marques, E.A.S., Barbosa, A.S.Q.F., Carbas, R.J.C., Akhavan-Safar, A., da Silva, L.F.M.: Introduction to Adhesive Bonding. John Wiley & Sons (2021)
7. E. Minin, M.F., Dilger, K.: Adhesive bonding of aluminum die cast components. n.d
8. European Aluminium Association. Adhesive bonding. In: Aluminium Automotive Manual–Joining, chapter 9. European Aluminium (2022)
9. Henkel. Loctite silane modified polymers (smp)–technology overview n.d
10. Petrie, E.M.: Ms polymers in "hybrid" sealants. Adhesive and Sealant Council Tech Paper (2010)
11. Li, W., Bouzidi, L., Narine, S.S.: Current research and development status and prospect of hot-melt adhesives: a review. Ind. Eng. Chem. Res. **47**(20), 7524–7532 (2008)
12. Gharde, S., Sharma, G., Kandasubramanian, B.: Hot-melt adhesives: Fundamentals, formulations, and applications: a critical review. Prog. Adhes. Adhes. **6**, 1–28 (2021)
13. Burggra¨f, P., Bergweiler, G., Kehrer, S., Krawczyk, T., Fiedler, F.: Mega-casting in the automotive production system: expert interview-based impact analysis of large-format aluminium high-pressure die-casting (hpdc) on the vehicle production. J. Manuf. Process. 124, 918–935 (2024)
14. Rodrigues, V.C.M.B., et al.: Mechanical characterization of a novel cyclic olefin-based hot-melt adhesive. Materials **18**(4), 855 (2025)
15. Bainbridge, I.F., Taylor, J.A.: The surface tension of pure aluminum and aluminum alloys. Metall. Mater. Trans. A **44**, 3901–3909 (2013)
16. Razzouk, E., Koncz-Horva´th, D.I., To¨r¨ok, T.: Critical challenges in the anodizing process of aluminium–silicon cast alloys–a review. Crystals **14**(7), 617 (2024)
17. ASTM International. Astm d6862: Standard test method for 90 degree peel resistance of adhesives. Technical report, ASTM International, West Conshohocken, PA (2021). https://doi.org/10.1520/D6862-11R21
18. Borges, C.S.P., et al.: Testing method to determine the strength and fracture toughness of adhesives in a single continuous test. Theor. Appl. Fract. Mech. **128**, 104119 (2023)

19. Kendall, K.: Thin-film peeling–the elastic term. J. Phys. D Appl. Phys. **8**(13), 1449 (1975)
20. Kaelble, D.H.: Theory and analysis of peel adhesion: rate-temperature dependence of viscoelastic interlayers. J. Colloid Sci. **19**(5), 413–424 (1964)
21. Creton, C., Ciccotti, M.: Fracture and adhesion of soft materials: a review. Rep. Prog. Phys. **79**(4), 046601 (2016)

Applications

Adhesive Technology in Shipbuilding–Opportunities and Requirements

L. Fröck[1(✉)], P. Kaeding[1], and N. Kayser[2]

[1] Faculty of Mechanical Engineering and Shipbuilding, University of Rostock, Albert-Einstein-Str. 2, Rostock, Germany
{linda.frock,patrick.kaeding}@uni-rostock.de
[2] Fraunhofer IGP, Albert-Einstein-Str. 30, Rostock, Germany
nele.kayser@igp.fraunhofer.de

Abstract. The national shipbuilding industry has undergone significant transformation over the past decades. While tankers and container ships were built in the 1990s, shipyards today primarily construct cruise ships and specialized vessels, making each project a unique, one-of-a-kind object. To remain competitive in the international market and achieve the shortest possible construction times, the overlap between the planning and manufacturing phases has continuously increased. This often results in late-stage design modifications and the need for additional components during advanced construction phases, where coatings and insulation may already be in place. Adhesive bonding technology offers a low-heat joining method that prevents damage to pre-installed components. However, designing such adhesive joints require both structural modifications to components according to the selected joining technique and the consideration of aging-related reduction factors. In the project "Development and Validation of Laboratory Aging Cycles for the Approval of Shipbuilding Adhesive Bonds," aging cycles for structural and highly elastic thick-layer adhesives were investigated for both interior and exterior ship applications. The study analyzed individual stress factors (e.g., humidity, temperature, and salt exposure) as well as combined mechanical-hygrothermal loading scenarios. Subsequently, the results were compared with data from real-world field aging tests. For the adhesives used, it is possible to estimate degradation factors, enabling structural modifications of components such as stiffeners or reinforcing plates to ensure sufficient adhesive bonding areas. These components, which are typically welded, cannot be bonded in their existing form, as proper bonding surfaces must be created to effectively transfer loads. Therefore, structural adhesive bonding is only feasible with adapted designs and a detailed load-bearing capacity assessment over the ship's service life.

Keywords: Adhesive bonding · shipbuilding · aging · ship structures

L. F. M. da Silva and R. D. Adams (Eds.): AB 2025, PEM, pp. 173–185, 2026.
https://doi.org/10.1007/978-3-032-19022-2_10

1 Introduction

In shipbuilding, as in the transport sector in general, there is an increasing use of novel and innovative materials [1–3]. This development is driven, on the one hand, by the higher proportion of lightweight structures and, on the other hand, by the growing demands regarding the aesthetics of the manufactured products. As a result, there is a need to apply modern joining techniques, such as adhesive bonding, in order to join both dissimilar and similar materials with minimal heat input and without the necessity of pre-drilling.

2 State of the Art

2.1 Adhesive Bonding in Shipbuilding

The research project *Bondship* [4] (2000–2003) was the first to develop systematic approaches addressing various aspects of adhesive bonding in shipbuilding. This included methods for selecting suitable adhesive systems, designing and analyzing bonded joints, testing materials and structures, performing non-destructive evaluation, and producing bonded connections. The project results established an important foundation for the application of adhesive bonding technology in shipbuilding. However, no methodology was defined within this framework for accelerated aging of adhesives and bonded joints, leaving this aspect unaddressed.

The project *Standardization of Highly Elastic Aluminum–Steel and FRP Bonded Joints for Shipbuilding* already demonstrated the feasibility of bonding deckhouse structures as early as 2014 [5]. Further projects, such as *Adhesive Bonding of Piping Systems in Shipbuilding* and *Rework-Free Bonding of Mounting Brackets on Ship and Steel Structure Coatings*, provided evidence for the applicability of bonding technology during the outfitting phase of ships [6, 7].

The use of adhesive bonding in shipbuilding, due to the high safety requirements of the sector, requires approval by a classification society. Because of the lack or insufficiency of established guidelines from the classification societies, such approvals are typically granted on a case-by-case basis. To obtain such approval, the load-bearing capacity of the bond must be demonstrated. This involves, on the one hand, providing evidence of adhesive-compliant design of the components and joints and, on the other hand, verifying the strength calculation including reduction factors.

To determine a reduction factor accounting for aging under conditions typical of ship service, accelerated aging tests are necessary. While such protocols are standardized in automotive engineering, no defined aging cycles exist for shipbuilding. This significantly complicates the use of adhesive bonding and in many cases prevents it altogether. To address this, various research projects have been and are being conducted to investigate aging processes and adhesive degradation. One example is the AiF cluster project *BestKleb*, which was completed in 2015. The objective of this project was to understand the chemical processes occurring during the aging of bonded joints and to develop and evaluate approaches for service life prediction [8, 9]. The investigations applied artificial aging conditions, such as humid air or water immersion at 60 °C. The results provided valuable insights into the mechanisms of adhesive aging and their impact on the mechanical properties of bonded joints. However, no reference to shipbuilding was established, since tests under maritime conditions—such as saltwater, salt spray, or seawater immersion—were not performed. For this reason, the project *Development and Validation of Laboratory Aging Cycles for the Approval of Shipbuilding Bonded Joints* identified accelerated laboratory aging procedures for exterior and interior applications on the basis of field aging data.

2.2 Aging of Adhesive Joints for Use in Maritime Structures

In ship operation, a variety of environmental loads occur. For the structural adhesive joints presented in this article, the most relevant factors are moisture, salt exposure, and temperature. In the case of highly elastic thick-layer adhesives, UV radiation can also be relevant depending on the design; however, this will not be discussed further here. The following paragraphs briefly describe the fundamental aging effects caused by the three primary loads.

Water ingress into the adhesive layer can trigger various effects. Swelling of the adhesive, for which studies [10, 11] demonstrated a linear correlation with water content, leads to internal stresses within the bonded structure [12]. Significant changes in mechanical properties can also occur, such as plasticization, without permanent damage to the adhesive [13]. Over longer periods, the adhesive can degrade via hydrolysis or leaching of specific constituents (erosion), processes which occur more slowly than the reversible effects [14–16]. Additionally, water affects the adherends—for example, by swelling, changes in viscoelastic properties, or corrosion at the interfaces (see Fig. 1) [17].

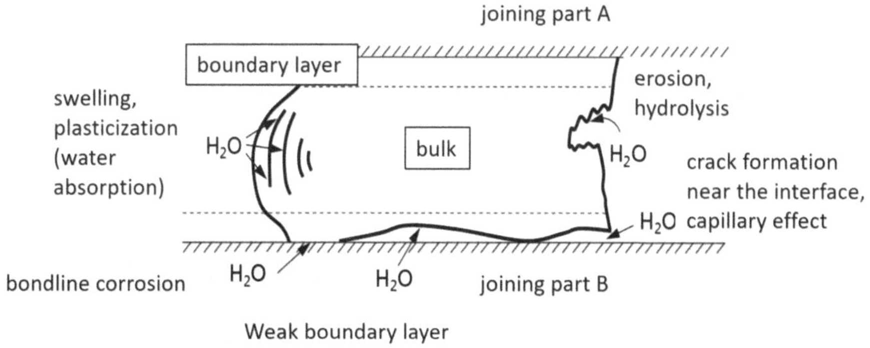

Fig. 1. Schematic representation of the mechanisms of action of water on the adhesive bond according to [18]

Salts in combination with moisture act as particularly aggressive corrosive agents on metallic bonded joints. The presence of salts increases the conductivity of water, creating an electrolyte that accelerates the release of metal ions from the metallic adherends and thus leads to corrosion damage [19]. The availability of oxygen is a critical factor. Chloride ions diffuse into the bond line from the joint edges and induce interfacial corrosion (bondline corrosion). This corrosion propagates from the affected area at a constant rate. The susceptibility of the bonded joint largely depends on the corrosion resistance of the adherend materials. For example, bonds involving low-alloy steels are more prone to degradation than those made from high-alloy steels or seawater-resistant aluminum alloys [20, 21]. Additionally, adhesives containing metallic particles can increase conductivity between two metallic adherends, thereby promoting corrosion [19].

3 Experimental Investigations

3.1 Investigation of Thermal Loads

In the first step, the different loads were examined in isolation as far as possible in order to assess their criticality and the associated damage mechanisms. As specimen types, both lap shear specimens in accordance with DIN EN 1465:2009-07 and tensile specimens per DIN EN ISO 527-2:2012-06 Type 1BA were prepared using two highly elastic thick-layer adhesives and two structural adhesives. However, this work will subsequently focus only on the lap shear specimens bonded with structural adhesives. The adherend materials were selected to reflect shipbuilding applications, where steel and fiber-reinforced composites (e.g., steel hulls and FRP superstructures) are bonded. Due to the large overlap lengths typically required in shipbuilding to transfer loads, a double overlap length compared to DIN EN 1465:2009-07 was selected for the specimens. The adhesive layer thickness reflects the high tolerances commonly encountered in shipbuilding. The specimen parameters are summarized in Table 1. These specimens were used identically for all subsequent tests.

Table 1. Parameters of the lap shear specimens used

Parameter	Adherend 1	Adherend 2	Lap shear specimens
Material	*GFK*	*V4A*	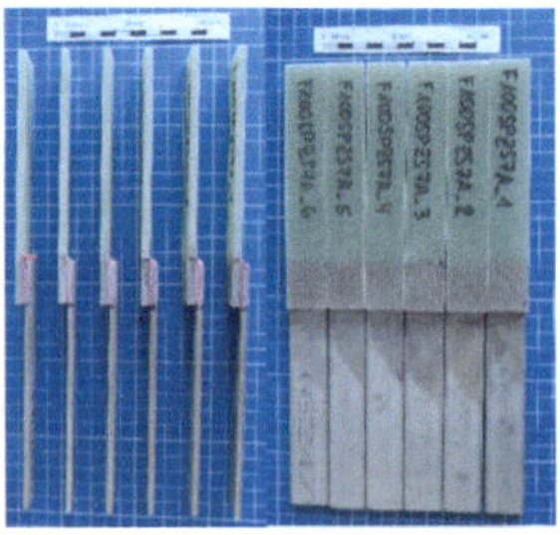
Length *[mm]*	*150*	*150*	
Width *[mm]*	*25*	*25*	
Thickness *[mm]*	*5*	*4*	
Overlap length *[mm]*	*25*		
Adhesive layer thickness *[mm]*	*1*		

After adhesive curing, the specimens were subjected to various laboratory aging exposures. To assess the influence of temperature on bond strength, three different exposures lasting 3, 6, and 12 weeks each were performed. Two constant temperature exposures (55 °C and 80 °C) and one cyclic exposure (0–55 °C) were selected based on typical service temperatures in shipbuilding. As previous projects had shown that negative temperatures have a comparatively minor effect, they were not considered here. The exposure parameters are provided in Table 2.

Table 2. Parameters of thermal exposure

Exposure Name	Max. Temperature [° C]	Min. Temperature [° C]	Exposure Cycle
T55–10	*55*	*10*	*0,5 (12 h + 55 °C, 12 h 0 °C)*
T55	*55*	*55*	*1* (constant exposure)
T80	*80*	*80*	*1* (constant exposure)
Exposure duration per series: 3, 6, and 12 weeks			

After thermal exposure, all specimens were conditioned for at least 72 h under standard laboratory climate [22] and subsequently tested in lap shear loading. The test parameters were kept constant for all subsequent experiments and are given in Table 3.

Table 3. Parameters of the lap shear tests

Testing machine	*Zwick Z400*
Testing speed	1 mm/min, displacement-controlled
Preload	none
Test climate	Room temperature, uncontrolled humidity
Load direction	Lap shear (load parallel to adhesive layer)
Termination criterion	*70% Fmax*
Specimen conditioning	*24h @ 23 °C/50% RH*

The lap shear strength results show that Structural Adhesive 1 exhibited strength reductions under exposures T55 and T55-10. No clear time-dependent trend was observed in the measured values (cf. Figure 2). In contrast, Structural Adhesive 2 exhibited a reduction in strength under T80 exposure, with increasing exposure duration resulting in a greater reduction. This effect was not observed for the other temperature conditions.

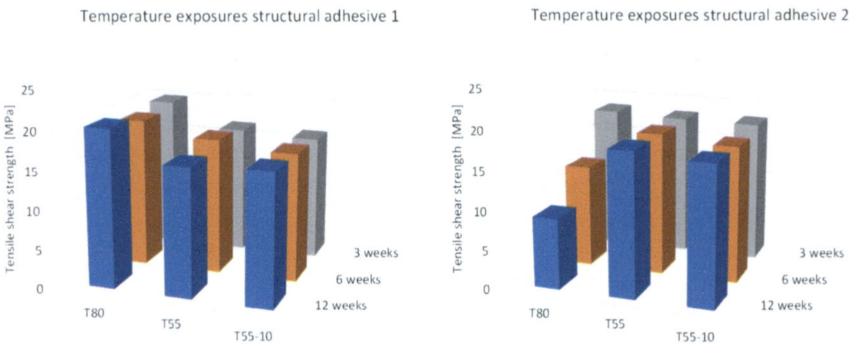

Fig. 2. Lap shear strength after different temperature exposures and durations

3.2 Investigation of Moisture-Induced Effects

To investigate the influence of moisture, the same specimens were subjected to three different humidity exposures. In addition to the ambient humidity (60% relative humidity up to immersion in water), the exposure type was varied between 12 weeks of continuous exposure and 24 weeks of cyclic exposure (7 days exposed + 7 days unexposed). Moreover, the impact of subsequent standard climate conditioning for 7 days compared to 1 day was examined. The specimens were again destructively tested. Figure 3 shows that Structural Adhesive 1 exhibited only minor reductions in strength. Due to limited availability of Structural Adhesive 2, two series in the test plan were omitted. Structural Adhesive 2 showed much greater sensitivity to the exposures, and cyclic exposure resulted in less strength reduction.

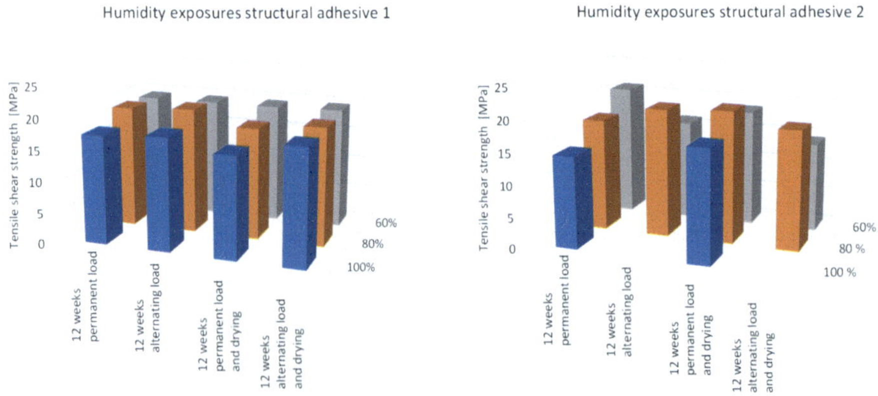

Fig. 3. Lap shear strength after different humidity exposures and loading types

3.3 Investigation of the Influence of Varying Salt Content in Salt Spray Tests

In this project, three different salt concentrations were investigated. Concentrations of 1% (e.g. in the Baltic Sea) and 5% were selected to reflect realistic seawater salinities (cf.

Figure 4). Additionally, a 10% salt concentration was included to deliberately increase the severity of exposure and accelerate aging effects. The exposures were conducted at a constant temperature of 35 °C. Exposure durations were 2, 5 and 10 weeks.

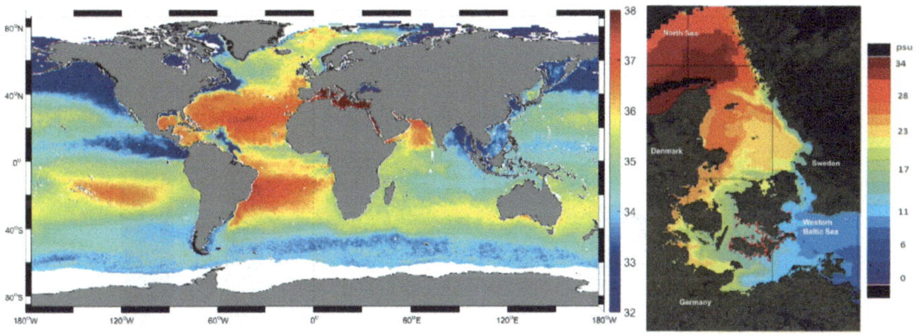

Fig. 4. Left: Global sea-surface salinity of the world's oceans [23], Right: Salinity data (yearly mean) in the Baltic Sea-North Sea [24] in Practical Salinity Unit (psu)

The lap shear strength results for the structural adhesives indicate that salt spray exposure did not have a significant influence on the bond strength of the adhesive joints. However, studies from other projects have shown that field exposure in seawater can lead to higher reduction factors than the water immersion tests in this project [25]. It is therefore assumed that the sodium chloride solution does not replicate the full effect of real seawater electrolytes on bonded joints. Further investigations into this issue are necessary.

3.4 Field Exposure and Accelerated Laboratory Aging

In parallel to the investigation of individual parameters, field exposure of specimens was conducted on the pilot station vessel *Elbe* in the Elbe estuary, both inside and outside the ship (cf. Fig. 5).

Fig. 5. Field exposure of specimens on the pilot station vessel Elbe (left), attached to the outer (middle) and inner (right) part of the ship

The lap shear strength results show that in the interior environment, hardly any changes in strength occurred for the two structural adhesives over time. In contrast, in the exterior environment, strength reductions of approx. 7% were observed for Structural Adhesive 1 after 24 months compared to the unaged reference measurements (cf. Fig. 6). The specimens bonded with Structural Adhesive 2 could not be tested because they were damaged. This was due to the use of the adhesive with a 1 mm bond line thickness for which it was not designed.

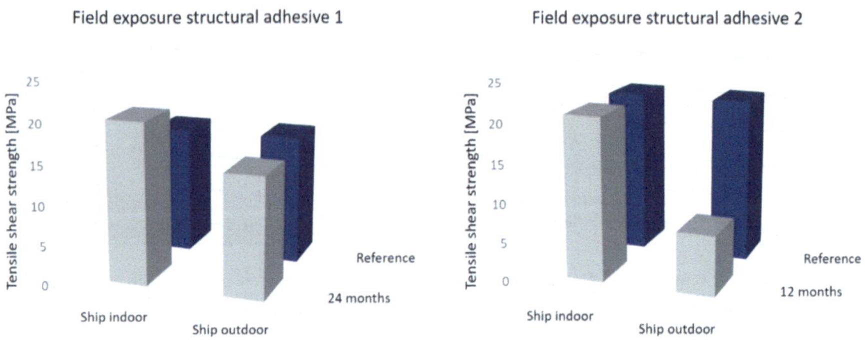

Fig. 6. Lap shear strength after aging of structural adhesives in interior and exterior ship environments for 12 and 24 months

The corresponding fracture patterns of structural adhesive 1 are shown in Fig. 7.

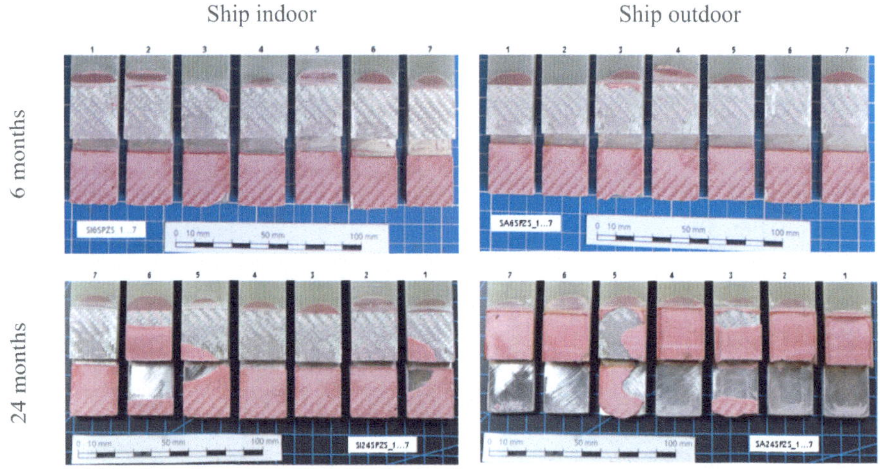

Fig. 7. Fracture patterns of structural adhesive 1 for the exposures ship indoor and ship outdoor

From the experiments, reduction factors were calculated and subsequently used for the design of bonded joints. For Structural Adhesive 1, a reduction factor of 0.93 was determined. Due to the limited exposure duration, this value applies to a service period of 2 years. Specimens are currently being stored on the ship for an additional 5 years to determine longer-term reduction factors. To enable faster determination of reduction factors for different adhesives in the future, various accelerated laboratory aging protocols were conducted in parallel. The best correlation with field aging results (strength and fracture morphology) was achieved for the structural adhesives in exterior environments with combined temperature–humidity exposures, and for interior environments with purely thermal exposures (see Table 4). As many shipbuilding classification societies require salt spray exposure, this can optionally be applied prior to other aging steps.

Table 4. Accelerated laboratory aging protocols for simulating 2 years of real aging in ship exterior and interior environments

Environment		Structural Adhesives
Exterior	*Variant 1*	5 weeks temperature + humidity: 80 °C/80% RH
	Variant 2	10 weeks temperature + humidity: 55 °C/80% RH
Interior	*Variate 1*	12 weeks temperature: 55 °C
	Variant 2	3 weeks cyclic exposure: 10 °C/55 °C

4 Incorporation of Reduction Factors into Structural Design and Outlook

Compared to traditionally welded constructions, adhesive bonding solutions must be adjusted in their joint design. To enable a safe and efficient design of adhesive joints in practical construction, reduction factors that represent the strength of the adhesive bond at the end of the structure's service life are a crucial contribution. Ongoing aging studies will progressively provide these factors. With appropriately dimensioned and aged substructures, such as typical stiffened plates in shipbuilding, calculations and tests for load-bearing capacity can then be conducted, as is customary when considering the structural safety of steel structures [26, 27].

A literature survey shows that the regulations of classification societies specify requirements for design [28], while the actual implementation is often based on shipyard traditions that are only partially documented. [29] investigates optimization possibilities for existing welded steel structures. [30] highlights potential adaptations for integrating storage systems for fossil-free energy carriers in ships. An overview of construction methods, calculation and dimensioning techniques, as well as alternative materials and joining technologies, can be found in [31]. Application-oriented studies on the design, manufacturing, and testing of bonded metallic structures in maritime environments are documented in [4], though they focus primarily on local implementations. [32] illustrates adhesive joining using the example of offshore wind energy installations.

To integrate reduction factors due to aging into the structural design–along with other requirements–, the application of probabilistic verification concepts to bonded maritime large structures enables a sound safety assessment and the development of innovative, reliable, and sustainable construction methods. Systematization of superimposed loads and their effects across size scales for assessing the reliability of safety-critical, highly stressed (FRP) structures is presented in [33]. This methodology is applied in [34] to an aircraft with an integrated H_2 storage system in order to analyse load cases, failure mechanisms, and optimization potentials. Both studies employ Bayesian networks to model error and event chains of safety relevance.

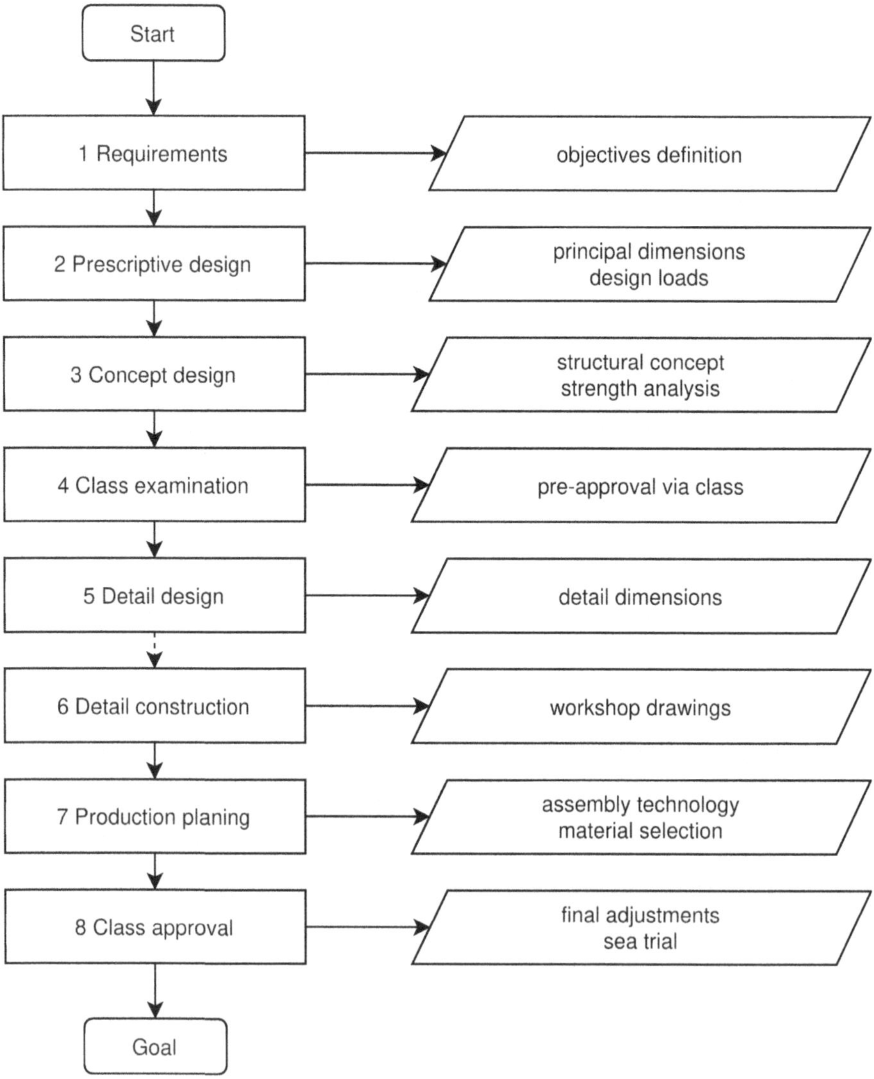

Fig. 8. Structural design process of a ship

Drawing on established procedures from classification societies and grounded in the physical-mathematical relationships among functional, structural, material, and adhesive parameters–as well as damage and failure modes–future work will develop a tailored methodology for the adhesive-compatible design, dimensioning, and verification of large-scale maritime structures. It will particularly focus on phase 3 "concept design" of the structural design process of a ship, Fig. 8, to develop structures that are tailored for adhesive bonding.

5 Conclusion

This work demonstrates that adhesive bonding is a viable and promising joining technology for maritime structures, offering advantages in design flexibility and lightweight construction. However, its successful implementation requires a comprehensive understanding of environmental aging effects and their impact on joint performance. The conducted laboratory and field investigations enabled the derivation of reduction factors to account for moisture, temperature and salt exposure. This identified accelerated aging protocols that reliably simulate real conditions. By integrating these findings into structural design and adopting probabilistic verification concepts, adhesive bonding can be applied safely and sustainably in shipbuilding. Future research will focus on refining these methodologies and embedding them systematically into the early design phases to facilitate broader industrial acceptance and approval by classification societies.

References

1. Thiru, S. et al.: Enhancing marine composite performance: the role of Cu, Al, and graphene nanofillers in seawater-aged glass fiber epoxy composites. J. Mater. Res. Technol. **36**, 9713–9725 (2025). ISSN 2238-7854. https://doi.org/10.1016/j.jmrt.2025.05.204
2. He, L., Ye, Z.: Research on the performance of carbon fiber composite railings for ships. Acad. J. Sci. Technol. **13**(2), 158–164 (2024). https://doi.org/10.54097/2vd3zh14
3. Mayland, W.: Untersuchungen zum aktuellen internationalen Stand der Technik bezüglich Entwicklung und Anwendung von Klebeverbindungen in der Schiffsfertigung (2020)
4. Weitzenbock, J., McGeorge, R.: Bondship Project Guidelines (2005). ISBN 82-515-0305-1
5. Siering, K., Schneider, B., Klapp, O.: Schlussbericht zum IGF-Vorhaben 16764N "Standardisierung von hochelastischen Aluminium-Stahl-und FVK-Stahl-Klebverbindungen für den Schiffbau", Hamburg, Bremen (2014)
6. Fröck, L., Nagel, C.: Schlussbericht zum IGF-Vorhaben Nr. 17808 BG "Klebtechnisches Fügen von Rohrleitungen im Schiffbau", Rostock, Bremen (2015)
7. Fröck, L., Kordy, H.: Schlussbericht zum IGF-Vorhaben Nr. 18527 BG "Nachbearbeitungsfreies Halterkleben auf Schiff-und Stahlbaubeschichtungen" Rostock, Bremen (2018)
8. Possart, W., Brede, M.: Beständigkeit von Klebungen verstehen und berechnen (BestKleb), Tagungsband, 16. Kolloquium Gemeinsame Forschung in der Klebtechnik", Köln (2016)
9. Possart, W., Brede, M., eds.: Adhesive joints: ageing and durability of epoxies and polyurethanes. Wiley-VCH (2018)
10. Liljedahl, C.D.M., et al.: Modelling the environmental degradation of adhesively bonded aluminium and composite joints using a CZM approach. Int. J. Adhes. Adhes. **27**(6), 505–518 (2007)

11. Walter, H., et al.: In-situ-characterization of moisture induced swelling behaviour of microelectronic relevant polymers. 2012 13th International Thermal, Mechanical and Multi-Physics Simulation and Experiments in Microelectronics and Microsystems. IEEE (2012)
12. Heshmati, M.: Hygrothermal durability of adhesively bonded FRP/steel joints. Chalmers University of Technology, Gothenburg; Swe-den (2015)
13. Brede, M., et al.: On the energy release rate of aged adhesive joints. Adhesive Joints: Ageing and Durability of Epoxies and Polyurethanes, pp. 375–404 (2018)
14. Xiao, G.Z., Delamar, M., Shanahan, M.E.R.: Irreversible interactions between water and DGEBA/DDA epoxy resin during hygrothermal aging. J. Appl. Polym. Sci. **65**(3), 449–458 (1997)
15. Depollier, L., Huacuja-Sánchez, J.E., Possart, W.: Joints with a basic epoxy adhesive: ageing processes. Adhesive Joints: Ageing and Durability of Epoxies and Polyurethanes pp. 265–307 (2018)
16. Huacuja-Sánchez, J.E., Engel, P., Possart, W.: Steel joints with a basic polyurethane adhesive–ageing processes. Adhesive Joints: Ageing and Durability of Epoxies and Polyurethanes, pp. 309–353 (2018)
17. Seim, W., Eisenhut, L.: Long-term behaviour of adhesively bonded timber–concrete composites. Adhesive Joints: Ageing and Durability of Epoxies and Polyurethanes pp. 449–469 (2018)
18. Waite, J.H.: Nature‘s underwater adhesive specialist. University of Delaware, USA (1987)
19. Wendler-Kalsch, E., Gräfen, H.: Korrosionsschadenkunde, Klassiker der Technik, Springer-Verlag Berlin Heidelberg S.15 (1998). https://doi.org/10.1007/978-3-642-30431-6_2
20. Habenicht, G.: Kleben-Grundlagen, p. 797. Anwendungen, Springer Verlag, Berlin Heidelberg, Technologien (2009)
21. Rasche, M.: Handbuch Klebtechnik, Hanser Verlag, München Wien (2012). Print ISBN: 978-3-446-42402-9
22. DIN EN ISO 291:2008-08: Kunststoffe-Normalklimate für Konditionierung und Prüfung
23. European Space Agency–ESA/Climate Change Initiative (CCI), Global sea-surface salinity (2019). https://www.esa.int/ESA_Multimedia/Images/2019/05/Global_sea-surface_salinity. Accessed 19 Aug 2025
24. Behrens, J., Deurs, M., Christensen, E.: Evaluating dispersal potential of an invasive fish by the use of aerobic scope and osmoregulation capacity. PLoS ONE **12**, e0176038 (2017). https://doi.org/10.1371/journal.pone.0176038
25. Gatzke, J., Lemmrich, L., Fröck, L., Knape, A., Myslicki, S.: Steigerung der Festigkeit und Dauerhaftigkeit von Stahl-Klebungen im Offshorebereich, IGF-Vorhaben Nr. 20836 BG (2023)
26. Lindemann, T., Kaeding, P.: Investigations on the influence of shear and lateral loads on the collapse behaviour of plate structures under inplane thrust. Ship Technol. Res. Taylor Francis **66**(1), 38–56 (2019)
27. Lindemann, T., La Ferlita, A., Di Nardo, E., Kaeding, P.: Application of different methods to determine the ultimate strength of ships in bending. Proceedings of the OMAE 2023 International Conference on Ocean, Offshore & Arctic Engineering, Mel-bourne, Australia, OMAE2023-103731 (2023)
28. DNV Rules for Classification of Ships (2016). https://rules.dnv.com/servicedocuments/
29. Lindemann, T., Kaeding, P.: An approach to optimization in ship structural design using finite element and optimization techniques. In: ISOPE International Ocean and Polar Engineering Conference (pp. ISOPE-I). ISOPE (2010)
30. Aider, C.A., Roß, L., Ehlers, S., Kaeding, P., Lindemann, T.: Integration of a tank storage solution for alternative fuels on a RoRo ship. In: Advances in the Analysis and Design of Marine Structures (pp. 761–771). CRC Press (2023)

31. Xiaozhi Wang, X., Neil Pegg, N. (eds.): Committee V.3 materials and fabrication technology. Proceedings of the 21st International Ship and Offshore Structures Congress (ISSC 2022) (2022)
32. Albiez, M., Damm, J., Ummenhofer, T., Kaufmann, M., Vallée, T., Myslicki, S.: Hybrid joining of jacket structures for offshore wind turbines–Determination of requirements and adhesive characterisation. Eng. Struct. **259**, 114186 (2022)
33. Schlegel, D., Vater, M., Spitzer, S., Gude, M., Hurtado, A.: Multi scale systematisation of damage and failure modes of high-pressure hydrogen composite vessels in aviation, Part 1: methodology. Int. J. Hydrogen Energy (2024). https://doi.org/10.1016/j.ijhydene.2024.11.115
34. Schlegel, D., Vater, M., Spitzer, S., Gude, M., Hurtado, A.: Multi scale systematisation of damage and failure modes of high-pressure hydrogen composite vessels in aviation, Part 2: analysis. Int. J. Hydrogen Energy (2024). https://doi.org/10.1016/j.ijhydene.2024.11.457

Microstructure and Interfacial Transition Zone in Recycled Aggregate Concrete: A Review of Old Concrete Waste Utilization

Seweryn Malazdrewicz(✉)

Wroclaw University of Science and Technology, Wybrzeże Wyspiańskiego 27, 50-370 Wroclaw, Poland

seweryn.malazdrewicz@pwr.edu.pl

Abstract. The closed-loop economy in construction focuses on the continuous reuse of products and raw materials, with the main goals of saving resources and reducing waste. The recycling of building materials is an important factor. The guiding principle is cradle-to-cradle, where, in an endless cycle, 100% of all raw materials and materials, after being consumed, are returned to the biological and technical raw material cycle and reused. Although the potential of recyclates is recognised, recycling of building materials into concrete is still used rarely. One of the reasons is the weakest part that comes with incorporating concrete waste-interfacial transition zone (ITZ), which is critical due to poor adhesion between the old and new cementitious phases. That is why this paper covers state-of-the-art and development paths for ITZ improvement in recycled aggregate concrete (RAC) as well as the methods used to improve the adherent mortar regions of recycled coarse aggregate (RCA) from old concrete waste.

Keywords: concrete waste · interfacial transition zone · recycled aggregate concrete · microstructure of concrete

1 Introduction

Construction and demolition waste (CDW) accounts for more than one-third of all waste generated in the EU [1]. While some components of CDW hold high resource value, and technologies for waste separation and recovery are well established, the recycling and material recovery rates vary significantly across the EU—from below 10% to over 90%. Although countries like France, Austria, Germany, and Spain have achieved high recycling rates, they still dispose of over 1 million tonnes of CDW mineral fraction annually (see Fig. 1) [2]. Notably, France has made significant progress in reducing disposal volumes over the past 14 years, whereas improvements in other major contributors have been relatively minor.

Concrete constitutes approximately 70% of CDW [3], making it the most significant mineral fraction. As the second most consumed substance in the world after water, global concrete consumption is estimated at around 30 billion tonnes per year. Concrete waste arises from various sources, including returned concrete from ready-mix trucks,

L. F. M. da Silva and R. D. Adams (Eds.): AB 2025, PEM, pp. 186–203, 2026.
https://doi.org/10.1007/978-3-032-19022-2_11

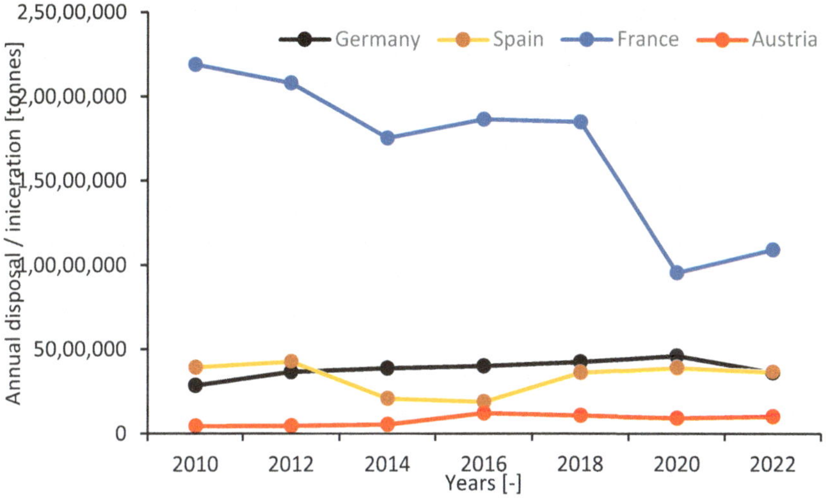

Fig. 1. Annual disposal / incineration in tonnes of selected countries: a) in years 2008 – 2022 b) in 2022 (Eurostat, 2025)

production waste from precast facilities, and CDW—the most significant contributor. Recycling concrete from CDW is crucial for conserving natural resources and promoting a closed-loop economy in construction. When properly separated before processing, the coarse fraction of recycled concrete aggregates (RCA) can meet quality standards for moderate incorporation in medium-performance concrete [4].

However, rather than being reused efficiently, concrete waste is often downcycled into secondary applications such as backfilling, noise barriers, road construction, and earthworks—suboptimal uses from both environmental and economic perspectives [5]. One of the primary barriers to the broader adoption of RCA in high-performance applications is its high variability in quality. Many recycling plants either lack the knowledge of optimal RCA production methods or are not incentivized to improve RCA quality for structural applications, leading to inconsistent final products [6].

Studies have shown that the properties of fresh and hardened recycled aggregate concrete (RAC) depend largely on the quality of the parent concrete used for RCA [7, 8]. The porosity of the source concrete and the crushing procedure are the two main factors influencing RCA characteristics [9]. This results in variations in density, water absorption, crushing value, and abrasion resistance compared to natural coarse aggregate (NCA) [10, 11]. A major challenge in using RCA for structural applications is the presence of adhered old cement paste, which creates multiple interfacial transition zones (ITZs) in RAC—between RCA and new mortar, NCA and new mortar, RCA and old mortar, and old mortar and new mortar. Since the ITZ is a critical phase that significantly affects RAC's mechanical properties, particularly due to weak adhesion at the interface between the recycled aggregate and the new cement paste, concerns persist regarding the structural safety and durability of RAC elements [12]. Nevertheless, research indicates that RAC made from high-quality parent concrete, such as high-strength concrete or precast elements, can achieve comparable or even superior performance to natural aggregate concrete [13]. Some national standards, such as those in China, Japan, and Denmark, permit the use of 100% RCA in concrete production when specific hardened-state properties are met and RCA quality is sufficiently high [14]. However, full RCA replacement remains rare [15]. In contrast, standards in Spain and the UK impose stricter limitations, allowing a maximum of 20% RCA replacement for concrete with strength class C40/50 [16].

With increasing research focusing on ITZ enhancement techniques—including carbonation treatment, mineral admixture incorporation, advanced mixing methods and improving adherent mortar zones —there is growing potential for improving RAC properties [17]. Given these advancements, there is a pressing need to consolidate current knowledge, identify existing research gaps, and outline future perspectives for optimizing RCA quality and ITZ performance.

2 Bibliometric Analysis

Given the increasing number of studies on ITZ RCA, a scientometric analysis of bibliometric data related to this topic was conducted to identify current research trends. This analysis visualizes input data and establishes connections between sources, keywords, authors, articles, and countries within the research domain. VOSviewer 1.6.20 was chosen for the scientometric analysis due to its ability to construct and visualize bibliometric networks based on journals, researchers, and individual publications [18]. It employs various analytical approaches, including citation analysis, bibliographic coupling, co-citation, and co-authorship relationships. The resulting maps illustrate key characteristics, their co-occurrences, and interconnections, offering valuable insights into the research landscape. To distinguish between different categories, specific colours are assigned to various elements in the mapping, while intensity mapping utilizes colour schemes such as "rainbow," "plasma," and "vibrios" to enhance data presentation and comprehension [19]. To import bibliometric data into the software, three databases were examined: Google Scholar, Scopus, and Web of Science. Scopus was ultimately selected as the primary database due to its higher number of publications compared to Web of Science and its superior quality control over Google Scholar [20]. According to its official website, Scopus provides 50%–230% more global content than its closest competitor, depending on the region [21]. The Scopus search yielded relevant results, and a filtering methodology was applied to exclude irrelevant materials. The process and selection criteria for preparing the visualizations are illustrated in Fig. 2.

The distribution of publications across subject areas is shown in Fig. 3. It illuminates the interdisciplinary nature of research on ITZ of RAC. Engineering leads with 41% of publications, underscoring the significant emphasis placed on mechanical properties. Materials Science follows with 33% highlighting the technological and material innovation focus. The presence of Environmental Science (7%), Physics and Astronomy (4%) and Energy (4%), further emphasizes the importance of environmental implications and sustainability aspects, physical and energy considerations in the development of RAC. Notably, the inclusion of diverse fields such as Computer Science, Business, Management and Accounting, Chemical Engineering, Earth and Planetary Sciences and Chemistry, albeit with lower percentages, indicates the broad potential impact of ITZ in RAC across various domains, underscoring the multifaceted approach to addressing sustainability in construction and material science. The analysis signals the critical intersections of environmental science, engineering, and materials science in advancing of ITZ analysis and its influence on RAC. However, inclusion of other areas serves as demonstration of possibilities for novel research inquiries, encouraging a holistic approach to addressing the sustainability challenges in the construction industry.

The document type analysis from Table 1a highlights the predominance of research articles, emphasizing the advancement of knowledge and innovation in ITZ and RAC. 44% of total publications in the field are among five prestigious with worldwide reach journals (Table 1b). It means innovative research concerning ITZ in RAC is of high importance and it serves as a benchmark for quality and relevance, aiding researchers in navigating the publication landscape effectively. Conference papers account for approximately 8% of total publications, indicating active discussions and knowledge exchange within the academic community, fostering collaboration and forward-thinking research approaches. Additionally, the presence of nine review papers suggests efforts to consolidate existing knowledge. However, a detailed analysis of these reviews in Table 2 reveals that ITZ is not their primary focus. This gap underscores the need for further studies aimed at identifying research gaps and outlining future directions on ITZ and its impact on RAC properties.

For the bibliometric analysis, a minimum keyword repetition threshold of five was applied in VOSviewer software. The network visualization in VOSviewer grouped keywords into clusters, each represented by a different colour. These clusters indicate the co-occurrence of specific keywords across various publications, with keywords (nodes) of the same colour belonging to related thematic groups. As a result, 143 out of 1756 keywords met the threshold. The 15 most frequently used keywords, along with their occurrences and total link strength, are presented in Table 3. These keywords can serve as a valuable guide for future authors to enhance the visibility and retrievability of their research. The dominant keyword themes revolve around mechanical properties, yet only compressive strength and tensile strength appear among the top 15. This suggests that future research should place greater emphasis on exploring how other properties of RAC, particularly durability, freeze-thaw resistance, and deterioration, are influenced by ITZ.

The analysis resulted in keywords visualisation, presented in Fig. 4. The network visualization in Fig. 4a represents keyword density and correlation frequency, indicated by node size and connecting paths, respectively. The more frequently a keyword appears, the larger its corresponding node. The results identified seven clusters of closely related publications:

- Cluster 1 (Red, 30 items): focuses on carbonation, microhardness, cracks, chloride penetration and compounds, numerical methods, and computer simulation.
- Cluster 2 (Green, 26 items): covers mechanical performance, permeability, density, hydrated lime, and silica.
- Cluster 3 (Blue, 26 items): includes self-compacting concrete, tensile strength, sustainable development, fibers, and mechanical and durability properties.
- Cluster 4 (Yellow, 24 items): relates to durability, porosity, frost resistance, and freeze-thaw cycles.

- Cluster 5 (Purple, 17 items): concerns demolition, high-performance concrete, cement, and concrete mixtures.
- Cluster 6 (Sky Blue, 13 items): focuses on recycled aggregates, fly ash, and geopolymers.
- Cluster 7 (Orange, 7 items): covers bending strength, particle size, and scanning electron microscopy.

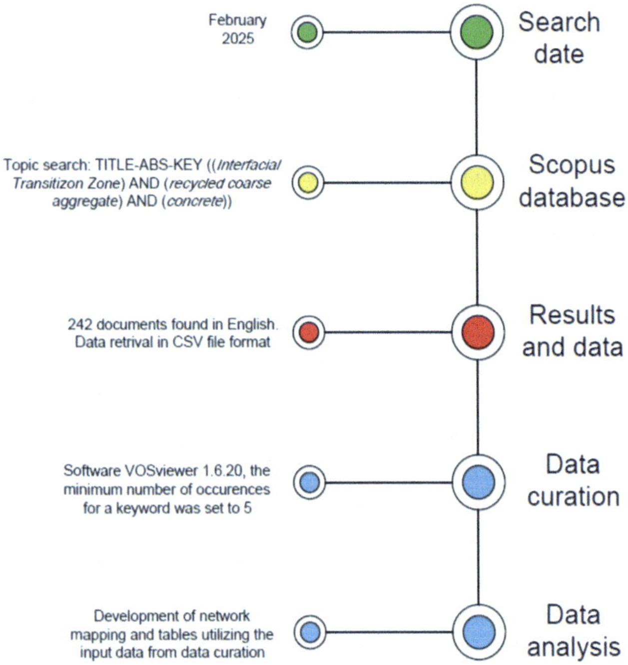

Fig. 2. The process of creating visualisation of keyword networks

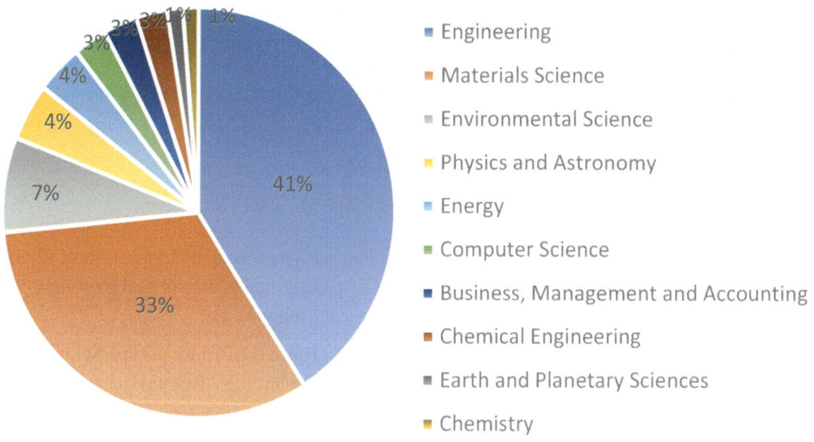

Fig. 3. Subject area spread and document categories of research articles

Table 1. Document categories and top five publication origins of analysed bibliometric data

Document Type	Amount of publications
Article	210
Conference paper	19
Review	9
Book chapter	3
Conference review	1

Journal Name	Amount of publications	Total number of citations
Construction and Building Materials	57	2797
Journal of Building Engineering	15	667
Journal of Cleaner Production	12	861
Journal of Materials in Civil Engineering	12	95
Materials	11	145

Table 2. Review publications among analysed bibliometric data

Name of review publication	Year	Journal
Progress in developing self-consolidating concrete (SCC) constituting recycled concrete aggregates: A review	2021	International Journal of Minerals, Metallurgy and Materials
Enhancing the quality of recycled coarse aggregates by different treatment techniques—a review	2021	Environmental Science and Pollution Research
Evaluation indexes of the frost resistance of recycled aggregate concrete and improvement mechanisms: A review	2024	Journal of Building Engineering

(continued)

Table 2. (*continued*)

Name of review publication	Year	Journal
Self-compacting concrete with recycled coarse aggregates from concrete construction and demolition waste – Current state-of-the art and perspectives	2023	Construction and Building Materials
Design and preparation methods for high-performance recycled aggregate concrete: a review	2024	Proceedings of Institution of Civil Engineers: Waste and Resource Management
The Impact of Fiber Reinforcement on Fresh and Physical Properties with Durability of Recycled Coarse Aggregate Concrete: A Review	2023	Journal of Engineering Science and Technology Review
Progress in the research of carbonation resistance of RAC	2020	Construction and Building Materials
Unraveling the Interplay of Physical-Chemical Factors Impacting the Carbonation Performance of Recycled Aggregate Concrete	2023	Materials
Effective recycling of surface modification aggregate using microwave heating	2014	Journal of Advanced Concrete Technology

Table 3. Fifteen most used keywords of analysed bibliometric data

No	Keyword	Occurrences	Total link strength
1	concrete aggregates	180	2074
2	recycling	166	1997
3	interfacial transition zone	136	1587
4	compressive strength	118	1435
5	recycled coarse aggregate	107	1216
6	recycled aggregate concrete	107	1103
7	aggregates	81	986
8	concretes	56	695
9	scanning electron microscopy	48	616
10	microstructure	48	614
11	mortar	52	598
12	cements	42	562
13	mechanical properties	46	558
14	water absorption	41	556
15	tensile strength	38	541

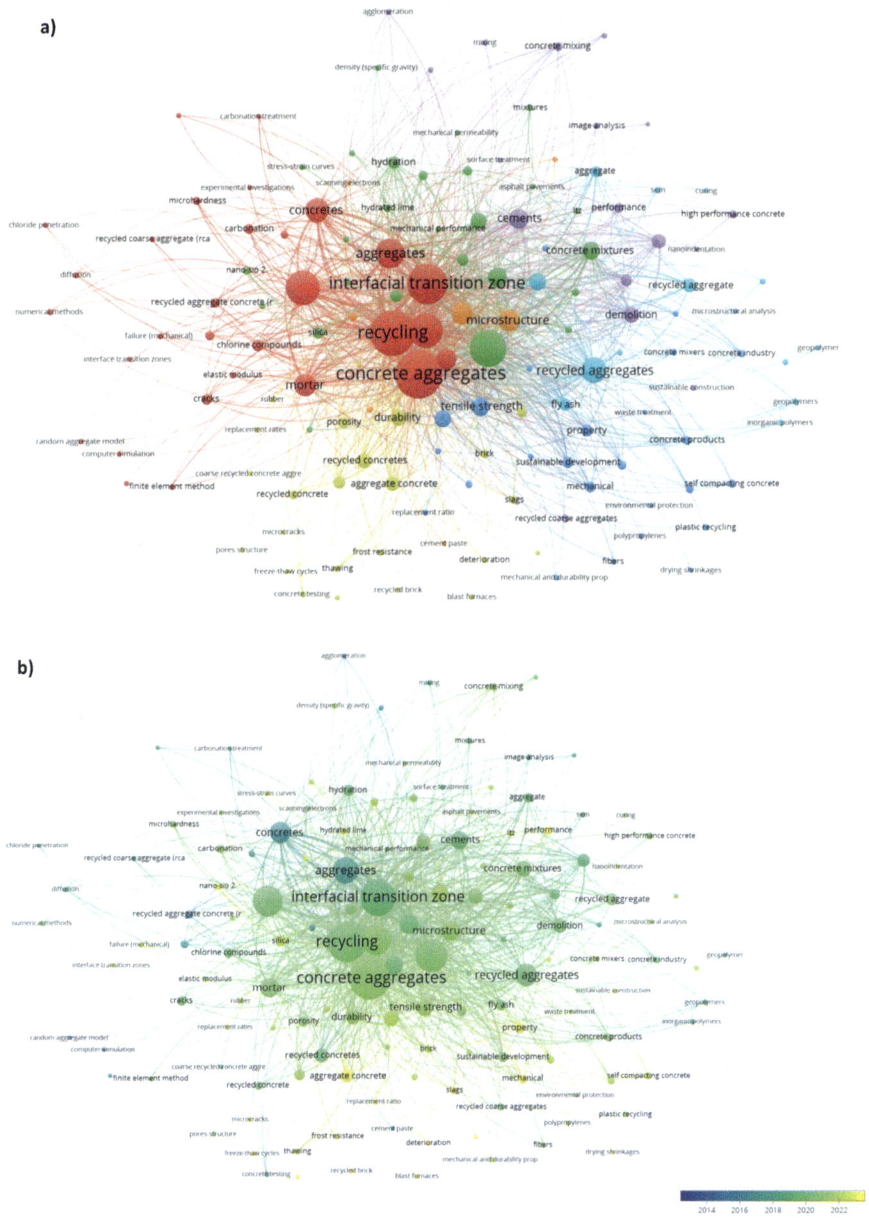

Fig. 4. Results of scientometric analysis: a) network visualisation b) density visualisation

3 Applications

ITZ is often considered the weakest part of concrete due to inadequate adhesion between aggregate and cement paste. It tends to lack large cement particles, which are unable to tightly pack around the aggregate, leading to a higher water-to-binder ratio compared to the bulk paste. The deep understanding of ITZ mechanisms and influence is need to enhance final cementitious composite properties. This section reviews existing knowledge on the ITZ in RAC, particularly parameters affecting ITZ and methods of its strengthening.

3.1 Main Parameters Affecting ITZ

RAC is a heterogenous material with a lot of variables affecting its uniformity. On the microstructural scale, starting from the relative movement of the paste and aggregate during the mixing process, there are many factors that can affect the microstructure of ITZ. Although there are many studies on ITZ influence on concrete, there is not much focus on parameters affecting ITZ. The formation of the ITZ microstructure mainly depends on the wall effect, which occurs near the larger aggregates, and the growth side effect, which relates to the distance from the aggregate surface. The summary of this section is presented in Fig. 5.

3.1.1 Mixing and Curing of Concrete

During the initial stage of hydration, cement particles exposed to water commence the hydration process, resulting in the formation of primary hydration products. Due to the uneven contact between cement particles and the aggregate surface, a microstructure with greater porosity develops within ITZ. The selection of appropriate mixing techniques can significantly influence the RCA microstructure during or at the onset of mixing. These techniques are classified based on the number of steps compared to conventional mixing (NM), primarily the two-step mixing (TSM) and the triple-mixing method (TM). The objective of employing suitable mixing methods is to produce a thin-coated cement slurry that fills pores and cracks while completing the mixing procedure effectively.

An example of TSM was demonstrated in [22], where the authors initially mixed fine and coarse aggregates for 60 s. Subsequently, half of the water and superplasticizers were added and mixed for an additional 60 s, followed by the introduction of cement and fly ash, which were mixed for another 30 s. Finally, the remaining water and superplasticizers were incorporated and mixed for 120 s. Silica fume was also introduced as a partial replacement for fly ash at varying percentages of 2%, 4%, 6%, 7%, and 8%. The findings indicated that the mixing process, along with the silica fume-cement slurry, effectively filled the pores and cracks present in ITZ. Furthermore, calcium hydroxide reacted with silicon dioxide, forming an additional binder similar to calcium silicate hydrate (C-S-H) crystals. The C-S-H gel produced from these reactions plays a crucial role in enhancing the strength, durability, and overall properties of the concrete. Similarly, [23] reported that the implementation of TSMA resulted in a denser old ITZ compared to the noticeable gaps observed in the microstructure of RAC prepared using NM.

3.1.2 Aggregate Type

Another significant factor influencing the strength of ITZ is the adhesion of cement paste to the aggregate's surface, which is largely determined by the roughness of RCA. Interestingly, [24] suggests that a rough surface and high local porosity allow water and cement paste to penetrate the aggregate, thereby forming a strong mechanical and physical bond between the cement paste and the aggregate. Conversely, [25] compared the pore structure of RCA with the new mortar in RAC. Their findings indicate that the adhesive (old) mortar attached to RCA, with its poor adhesive bonding to the new cement paste, exhibited the highest porosity defects among all ITZ bonds, which was identified as the primary factor contributing to the reduction of RAC compressive strength. Notably, research by [26] reported that concrete RCA contributes to higher minimum elastic modulus values than ITZs associated with organic components such as wood, plastic, and asphalt. The thickness of the RCA from concrete was measured at 55 μm, which was narrower compared to ITZ of asphalt (65 μm) and wood (60 μm) in cement paste, while ITZ of NCA and paste ranged from 10 to 50 μm. According to the authors, every effort should be made during RCA processing to minimize the presence of wood, plastic, glass, and asphalt in the final product due to their weak bonding with cementitious matrices. Study by [27] used high-quality precast rejects as RCA. Authors concluded that the additional interface between the attached mortar and the new cement paste is responsible for a better bond recycled aggregate/cement paste compared to natural aggregate/ cement paste.

3.1.3 Mineral Addition

The incorporation of mineral additions is a widely adopted strategy to enhance the properties especially of high-strength concrete. Mineral additions, including natural pulverized limestone and metakaolin, as well as industrial by-products such as granulated blast furnace slag, fly ash, and silica fume, improve particle packing density. This results in reduced porosity, a decreased w/c ratio and the densification of ITZ. Mentioned before study [22] has demonstrated that partially replacing fly ash with silica fume up to 8% effectively fills pores and cracks within ITZ through the formation of a silica fume-cement slurry. Furthermore, research [28] has shown that the incorporation of 10% silica fume can reduce ITZ thickness from 50 μm to 10 μm compared to concrete without mineral additions, leading to a twofold increase in interfacial bond strength, improved adhesion, and fracture energy. Additionally, it has been reported [29] that the inclusion of silica fume as a cement replacement significantly enhances interfacial bond strength, decreases the thickness of the near-surface weakness zone, and improves the ITZ within the near-surface layer of concrete. The authors identified 9% silica fume as the optimal content, providing substantial improvements in concrete strength. However, further increases beyond this level yield only marginal enhancements in the interfacial bond strength of near-surface-zone aggregates. Overall, the experimental findings indicate that the addition of mineral admixtures contributes to the formation of a denser ITZ and an optimized pore structure, resulting in enhanced mechanical performance [30].

3.1.4 ITZ Types

Another critical aspect is the composition of RCA itself, which features ITZs between the virgin aggregate and old matrix, as well as ITZs with the new matrix in various configurations, such as old matrix–new matrix and aggregate–new matrix. According to [25], for aggregate particle sizes ranging from 10 to 16 mm, the width of ITZ_4 (between aggregate and new matrix) exhibited an average elastic modulus and microhardness of approximately 15 GPa and 1, with a maximum width of 50 μm. This broad ITZ_4 adversely affects the homogeneity of RAC. Furthermore, the old ITZs, specifically the recycled aggregate-bonded mortar (ITZ_3) and new mortar-bonded mortar (ITZ_2), demonstrated even weaker mechanical properties, larger widths, and more pronounced interface effects. These characteristics significantly influence the performance of RAC, particularly in terms of crack development and overall structural integrity. Additionally, [31] investigated the differences between old and new ITZs in RAC. They found that the thicknesses of old and new ITZs are approximately 40–50 μm, increasing to 55–65 μm at a later age (90 days). The average indentation modulus of the old ITZ is 70–80% of that of the old paste matrix, whereas the average indentation modulus of the new ITZ is 80–90% of that of the new paste matrix. For the old ITZ in RAC, the lowest modulus occurs approximately 20–30 μm away from the natural aggregate surface. In contrast, for the new ITZ, the lowest modulus is located 10–20 μm away from the old paste aggregate surface. This distinction highlights the variation in mechanical performance and adhesive quality between the old and new ITZs and their contribution to the overall properties of RAC.

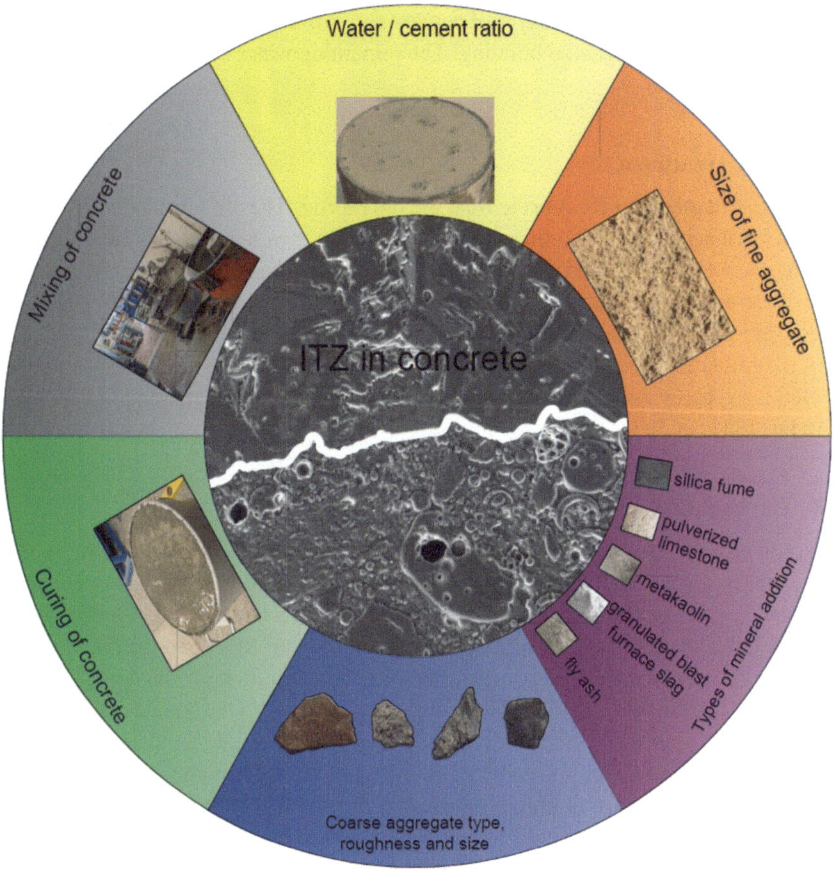

Fig. 5. Parameters directly and indirectly affecting ITZ in concrete

3.2 Methods of ITZ Improvement

According to [32], two key factors determine the durability of concrete: porosity and ITZ. Therefore, this section presents findings on ITZ enhancement and its impact on the properties of RAC. The focus is specifically on research that examines ITZ performance in RAC. Based on the literature review, the identified improvement methods have been categorized into five groups,

3.2.1 Mixing Approach

As highlighted in the previous section, the selection of appropriate mixing techniques can significantly influence the RCA microstructure. Usually the mixing approach are classified based on the number of steps compared to conventional mixing (NM), primarily the two-step mixing (TSM) and the triple-mixing (TM). These methods can differ between themselves in number of steps, mixing times, order of adding ingredients and the presence of admixtures and additions. The idea behind increased number of steps in mixing is to form a thin-coated layer of cement slurry that improves adhesion by sealing pores and enhancing interfacial bonding. The remaining water added later on is to finish the mixing procedure.

3.2.2 RCA Treatment

The key to strengthening RCAs is to treat the old mortar. The old mortar can not only be removed by mechanical grinding, ultrasonic cleaning, heating and chemical treatment but can also be strengthened by chemical, and calcium metasilicate slurry, wrapping methods, cement treatment, warm mix asphalt, hydrated lime, nanomaterials, super plasticizers, bio-deposition approach, silane-based water repellent, polymer solution [33]. Every treatment method enhances the quality of RCA by improving adhesion between recycled aggregates and the cement matrix, thus enhancing mechanical and durability properties [34].

3.2.3 Carbonation Treatment

The application of CO_2 curing presents a viable alternative to traditional RCA treatment. It takes a relatively long time but provides more effective improvement on the performance of recycled aggregate [35]. During the accelerated carbonation process, CO_2 penetrates the pores of the residual mortar adhered to RCA and chemically reacts with calcium hydroxide ($Ca(OH)_2$) and other cement hydration products present in the old mortar. This reaction results in the formation of calcium carbonates ($CaCO_3$) and calcium silicate hydrate (C-S-H) gels, which contribute to the densification of RCA by increasing the volume of the solid phase and filling existing pores and micro-cracks [36]. Consequently, carbonation treatment effectively reduces the porosity and water absorption capacity of RCA while enhancing ITZ of RAC without introducing impurities. Furthermore, this method offers an environmental advantage by capturing CO_2 from the industry [37], thereby contributing to reduced greenhouse gas emissions. This process demonstrates significant potential for promoting sustainability in civil engineering applications. However, several factors should be considered to optimise the mineralisation such as different carbonation pressures, duration, initial moisture content of RCA.

3.2.4 RCA Source

Most studies concerning RCA for concrete mixes, present laconic description of how RCA was acquired. This may suggest that each case concerning CDW to be used as RCA should be considered separately. However, literature review [13] and previous studies [38] show that precast elements, CDW from structures built with good quality of concrete are characterised without rapid decrease of properties for eco-concrete. It would deliver an opportunity for the precast concrete industry to reuse their debris without having to take exceptional treatment and without any notable lessening in the performance of concrete.

3.2.5 Mineral Addition

Previous studies have demonstrated that the incorporation of supplementary cementitious materials (SCMs) and natural pozzolanic materials can significantly enhance the performance of ITZ. The physical and chemical effects of SCMs, primarily the filler effect and pozzolanic reaction, contribute to microstructural refinement, leading to improved macro-scale performance [39]. Extensive research has been conducted on ITZ of RAC incorporating various SCMs, such as silica fume, steel slag powder, geopolymer binders, and fly ash [40, 41]. However, most studies focus on the influence of a single SCM on a single ITZ property, with limited research on the collective impact of multiple ITZs on the overall performance of RAC. Further investigation is needed to elucidate the mechanisms by which mineral additives modify multiple ITZ structures, reduce porosity, and optimize SCM combinations to enhance ITZ performance across different RAC types.

4 Discussion and Knowledge Gaps

Over the past decades, extensive research has focused on reducing the permeability and enhancing the durability of RAC [42–44]. As discussed before, these improvements are closely linked to strengthening ITZ. Despite advancements in treatment methods, several critical knowledge gaps remain:

1) Microstructural Characterization & Modelling

The nanoscale mechanisms governing ITZ formation, hydration product deterioration, and failure progression remain poorly understood [45]. The role of nanomaterials (e.g., nano-silica, graphene oxide) in ITZ modification requires further study. There is a lack of comprehensive models that accurately predict ITZ behaviour under different environmental and load conditions and mechanism of the failure progress [46, 47]. A reliable method to separate ITZ from the cement matrix and determine ITZ thickness is yet to be developed [48].

2) Long-Term Durability & Performance

More data is needed on how ITZ evolves over time due to environmental exposure, freeze-thaw cycles, and sulphate / chloride attacks [49]. The relationship between RCA replacement rate, long-term loading, and the degradation of RAC over a long time remains poorly understood [50].

3) Standardization of Testing Methods

The lack of uniform methodologies for ITZ assessment (e.g., thickness, porosity, mechanical properties) makes cross-study comparisons difficult. No standardized approach exists for evaluating RCA treatment efficiency across different recycled aggregate sources. There is a need for a unified methodology to assess ITZ characteristics.

4) Scaling Up for Practical Application

Research is often conducted on a laboratory scale, and the performance of ITZ-enhanced RAC in real-world construction scenarios is still uncertain. Therefore the cost-effectiveness and feasibility of ITZ improvement strategies at a large scale remain underexplored.

Addressing these gaps requires advanced imaging techniques, long-term durability studies, and standardized testing procedures. Future research should also leverage AI and machine learning to predict ITZ behaviour and optimize RAC mix designs for sustainable, high-performance concrete.

As shown in Fig. 4 researchers increasingly consider carbonation a promising approach for RCA treatment, with significant developments in this field since around 2020. Carbonation is particularly effective in reducing RAC permeability by directly enhancing ITZ properties. However, its effectiveness depends on factors such as the original cement composition of RCA [51] and the availability of a continuous CO_2 supply [52]. While carbonation holds great potential, further studies are needed to optimize its implementation across different RCA sources and construction scenarios. A comprehensive understanding especially on accelerated carbonation is required for its implementation in practice [53]. According to [35], strengthening RCA, particularly carbonation and slurry wrapping methods have the highest potential to be applied in practice because of their convenient operation, easy control and low cost. Overall, compared with slurry wrapping, carbonation treatment takes a relatively long time but provides more effective improvement on the performance of recycled aggregate. Therefore, based on the trend we might expect future researches improving carbonation and applying it in the industry.

5 Conclusions

Despite significant advancements in understanding the microstructure of RCA and its influence on concrete performance, rapid developments in material science have outpaced comprehensive reviews on ITZ in RAC. This review synthesized current research trends, identified key parameters affecting ITZ quality, and evaluated methods to enhance ITZ performance. Additionally, RCA improvement strategies were compared, and critical knowledge gaps were highlighted, providing a foundation for future research.

The main findings include:

- Concrete waste is predominantly downcycled into secondary applications due to quality variability, with full RCA replacement remaining rare. Some national standards impose strict 20% replacement limits for concrete with a strength class of C40/50. This review underscores the importance of sourcing RCA from prefabricated structures to ensure consistent and reliable material for structural applications.
- Scientometric analysis reveals increasing contributions from fields beyond engineering and material sciences, encouraging a more holistic approach to tackling sustainability challenges in construction.
- Parameters directly and indirectly having the greatest influence on ITZ—and thus on adhesion between aggregate and paste—in concrete were identified as mixing of concrete, water / cement ratio, size of fine aggregate, type of mineral addition, coarse aggregate type, roughness and size and curing conditions.

- Key gaps include microstructural characterization and modelling of ITZ, long-term durability assessment, standardization of testing methods, and large-scale application. Advanced imaging techniques, AI-driven predictive modelling, and standardized durability testing are crucial to bridging these gaps and optimizing RAC mix designs for high-performance, sustainable concrete.
- Researchers increasingly consider carbonation a promising approach for RCA treatment. Growing research interest in carbonation treatment suggests its potential to enhance RCA properties. Future studies should focus on optimizing carbonation processes for diverse RCA sources and construction scenarios to facilitate industrial implementation.

Funding. This research was funded from a project supported by the National Centre of Science, Poland [Grant no. 2020/39/O/ST8/01217 "Experimental evaluation of the fundamental properties of self-compacting concrete made of recycled coarse aggregate sourced from the demolition of large panel system buildings (RECSCCUE)"].

References

1. European Commission. Environment. (2022). https://environment.ec.europa.eu/. [Accessed 08 Aug 2025]
2. Eurostat. (2022). https://ec.europa.eu/eurostat. [Accessed 08 Aug 2025]
3. Akhtar, A., Sarmah, A.K.: Construction and demolition waste generation and properties of recycled aggregate concrete: a global perspective. J. Clean. Prod. **186**, 262–281 (2018)
4. Pacheco, J., De, B.J., Lamperti, T.M.: Use of recycled aggregates in concrete: opportunities for upscaling in Europe (2023)
5. Dettenborn, T., Harju, I., Forsman, J., Lehtonen, K., Ahlqvist, E., Kurki, T.: Recycling of concrete from demolished bridges and other engineered structures. IOP Conf. Ser. Earth Environ. Sci. **710**(1), 012064 (2021)
6. de Brito, J., Silva, R.: Current status on the use of recycled aggregates in concrete: where do we go from here? RILEM Tech. Lett. **1**, 1–5 (2016)
7. Gebremariam, H.G., Taye, S., Tarekegn, A.G.: Disparity in research findings on parent concrete strength effects on recycled aggregate quality as a challenge in aggregate recycling. Case Stud. Constr. Mater **19**, e02342 (2023)
8. Bhat, J.A.: Effect of strength of parent concrete on the mechanical properties of recycled aggregate concrete. Mater Today Proc. **42**, 1462–1469 (2021)
9. Gebremariam, H.G., Taye, S., Tarekegn, A.G.: Parent concrete strength effects on the quality of recycled aggregate concrete: a review. Heliyon (2024)
10. Li, L., Liu, Q., Huang, T., Peng, W.: Mineralization and utilization of CO2 in construction and demolition wastes recycling for building materials: a systematic review of recycled concrete aggregate and recycled hardened cement powder. Sep. Purif. Technol. **298**, 121512 (2022)
11. Malazdrewicz, S., et al.: Recycled coarse aggregate sourced from demolition of large panel system buildings: fundamental properties and perspectives for application. Arch. Civ. Mech. Eng. **25**(3), 1–23 (2025)
12. Li, W., Xiao, J., Sun, Z., Kawashima, S., Shah, S.P.: Interfacial transition zones in recycled aggregate concrete with different mixing approaches. Constr. Build. Mater. **35**, 1045–1055 (2012)
13. Malazdrewicz, S., Ostrowski, K.A., Sadowski, Ł: Self-compacting concrete with recycled coarse aggregates from concrete construction and demolition waste–Current state-of-the-art and perspectives. Constr. Build. Mater. **370**, 130702 (2023)

14. Plaza, P., del Bosque, I.S., Frías, M., de Rojas, M.S., Medina, C.: Use of recycled coarse and fine aggregates in structural eco-concretes. Physical and mechanical properties and CO2 emissions. Constr. Build. Mater. **285**, 122926 (2021)
15. Vintimilla, C., Etxeberria, M.: Limiting the maximum fine and coarse recycled aggregates-Type A used in structural concrete. Constr. Build. Mater. **380**, 131273 (2023)
16. Boletín Oficial del Estado (BOE). Structural Concrete Code. (2021). https://www.boe.es/eli/es/rd/2021/06/29/470. [Accessed 08 Aug 2025]
17. Memon, S.A., Bekzhanova, Z., Murzakarimova, A.: A review of improvement of interfacial transition zone and adherent mortar in recycled concrete aggregate. Buildings **12**(10), 1600 (2022)
18. VOSviewer. (2025). https://www.vosviewer.com/. [Accessed 08 Aug 2025]
19. Marathe, S., Sadowski, Ł, Shree, N.: Geopolymer and alkali-activated permeable concrete pavements: bibliometrics and systematic current state of the art review, applications, and perspectives. Constr. Build. Mater. **421**, 135586 (2024)
20. Harzing, A.W., Alakangas, S.: Google scholar, scopus and the web of science: a longitudinal and cross-disciplinary comparison. Scientometrics **106**(2), 787–804 (2016)
21. Scopus. (2025). https://www.scopus.com/. [Accessed 08 Aug 2025]
22. Rajhans, P., Panda, S.K., Nayak, S.: Sustainable self compacting concrete from C&D waste by improving the microstructures of concrete ITZ. Constr. Build. Mater. **163**, 557–570 (2018)
23. Tam, V.W., Gao, X.F., Tam, C.M.: Microstructural analysis of recycled aggregate concrete produced from two-stage mixing approach. Cem. Concr. Res. **35**(6), 1195–1204 (2005)
24. Zegardło, B., Szeląg, M., Ogrodnik, P.: Ultra-high strength concrete made with recycled aggregate from sanitary ceramic wastes–The method of production and the interfacial transition zone. Constr. Build. Mater. **122**, 736–742 (2016)
25. Lu, C., Yu, Q., Wei, J., et al.: Influence of interface transition zones (ITZ) and pore structure on the compressive strength of recycled aggregate concrete. Constr. Build. Mater. **456**, 139299 (2024)
26. Del Bosque, I.S., Zhu, W., Howind, T., et al.: Properties of interfacial transition zones (ITZs) in concrete containing recycled mixed aggregate. Cem. Concr. Compos. **81**, 25–34 (2017)
27. Salesa, Á., Pérez-Benedicto, J.Á., Esteban, L.M., Vicente-Vas, R., Orna-Carmona, M.: Physico-mechanical properties of multi-recycled self-compacting concrete prepared with precast concrete rejects. Constr. Build. Mater. **153**, 364–373 (2017)
28. Rossignolo, J.A.: Interfacial interactions in concretes with silica fume and SBR latex. Constr. Build. Mater. **23**(2), 817–821 (2009)
29. Xuan, D.X., Shui, Z.H., Wu, S.P.: Influence of silica fume on the interfacial bond between aggregate and matrix in near-surface layer of concrete. Constr. Build. Mater. **23**(7), 2631–2635 (2009)
30. Chen Q, Zhang J, Wang Z, et al. A review of the interfacial transition zones in concrete: Identification, physical characteristics, and mechanical properties. Eng. Fract. Mech. 109979 (2024)
31. Xiao, J., Li, W., Sun, Z., Lange, D.A., Shah, S.P.: Properties of interfacial transition zones in recycled aggregate concrete tested by nanoindentation. Cem. Concr. Compos. **37**, 276–292 (2013)
32. Fiol, F., Revilla-Cuesta, V., Thomas, C., Manso, J.M.: Self-compacting concrete containing coarse recycled precast-concrete aggregate and its durability in marine-environment-related tests. Constr. Build. Mater. **377**, 131084 (2023)
33. Raman, J.V.M., Ramasamy, V.: Various treatment techniques involved to enhance the recycled coarse aggregate in concrete: a review. Mater. Today Proc. **45**, 6356–6363 (2021)
34. Purushothaman, R., Amirthavalli, R.R., Karan, L.: Influence of treatment methods on the strength and performance characteristics of recycled aggregate concrete. J. Mater. Civ. Eng. **27**(5), 04014168 (2015)

35. Wang, J., Zhang, J., Cao, D., Dang, H., Ding, B.: Comparison of recycled aggregate treatment methods on the performance for recycled concrete. Constr. Build. Mater. **234**, 117366 (2020)
36. Poon, C.S., Shen, P., Jiang, Y., Ma, Z., Xuan, D.: Total recycling of concrete waste using accelerated carbonation: a review. Cem. Concr. Res. **173**, 107284 (2023)
37. Ang, P., Goh, W., Bu, J., Cheng, S.: Assessing carbon capture and carbonation in recycled concrete aggregates: a holistic life cycle assessment perspective with simulation at industrial scale. J. Clean. Prod. **474**, 143553 (2024)
38. Malazdrewicz, S., Ostrowski, K.A., Sadowski, Ł: Rheological and mechanical properties of self-compacting concrete with recycled coarse aggregate from the demolition of large panel system buildings. Appl. Rheol. **35**(1), 20240023 (2025)
39. Gao, S., Guo, X., Ban, S., Ma, Y., Yu, Q., Sui, S.: Influence of supplementary cementitious materials on ITZ characteristics of recycled concrete. Constr. Build. Mater. **363**, 129736 (2023)
40. Feng, W., Tang, Y., Zhang, Y., Qi, C., Ma, L., Li, L.: Partially fly ash and nano-silica incorporated recycled coarse aggregate based concrete: constitutive model and enhancement mechanism. J. Mater. Res. Technol. **17**, 192–210 (2022)
41. Ma, X., He, T., Xu, Y., Yang, R., Sun, Y.: Hydration reaction and compressive strength of small amount of silica fume on cement-fly ash matrix. Case Stud. Constr. Mater. **16**, e00989 (2022)
42. Levy, S.M., Helene, P.: Durability of recycled aggregates concrete: a safe way to sustainable development. Cem. Concr. Res. **34**(11), 1975–1980 (2004)
43. Thomas, C., Setién, J., Polanco, J., Alaejos, P., De Juan, M.S.: Durability of recycled aggregate concrete. Constr. Build. Mater. **40**, 1054–1065 (2013)
44. Xuan, D., Zhan, B., Poon, C.S.: Durability of recycled aggregate concrete prepared with carbonated recycled concrete aggregates. Cem. Concr. Compos. **84**, 214–221 (2017)
45. Guo, H., Wang, H., Xue, H., Li, H., Li, Y., Wei, L.: Study on damage deterioration mechanism and service life prediction of hybrid fibre concrete under different salt freezing conditions. Constr. Build. Mater. **435**, 136688 (2024)
46. Sun, D., Wu, K., Shi, H., Zhang, L., Zhang, L.: Effect of interfacial transition zone on the transport of sulfate ions in concrete. Constr. Build. Mater. **192**, 28–37 (2018)
47. Wang, R., Zhang, Q., Li, Y.: Deterioration of concrete under the coupling effects of freeze–thaw cycles and other actions: a review. Constr. Build. Mater. **319**, 126045 (2022)
48. Tu, Y., Yu, H., Ma, H., Liu, J., Li, L., Han, W.: Long-term study on the mechanical properties of the ITZ in coral aggregate concrete and its impact on material durability. Constr. Build. Mater. **449**, 138255 (2024)
49. Wang, Y., Wu, L., Wang, Y., Li, Q., Xiao, Z.: Prediction model of long-term chloride diffusion into plain concrete considering the effect of the heterogeneity of materials exposed to marine tidal zone. Constr. Build. Mater. **159**, 297–315 (2018)
50. Chen, B., Zhao, Y., Peng, L.: Long-term performance of recycled aggregate concrete beams exposed to 10 years of loading and chloride environments. Eng. Struct. **333**, 120140 (2025)
51. Masunaga, K.M., Tian, R., Iyoda, T.: Characterization of the Interfacial Transition Zone Between Carbonated Recycled Aggregates and New Cementitious Matrix: The Cement Type Influence on Carbonation. In: FIB Int Conf Concrete Sustainability, pp. 305–13 (2024)
52. Bao, H., et al.: Evolution of ITZ and its effect on the carbonation depth of concrete under supercritical CO2 condition. Cem. Concr. Compos. **126**, 104336 (2022)
53. Shuvo, A.K., Sarker, P.K., Shaikh, F.U.A.: Efficacy of various accelerated carbonation techniques to improve recycled concrete aggregates: a comprehensive review. J. Build. Eng. **95**, 110257 (2024)

Recycled Coarse Aggregate from Demolished Prefabricated Buildings in Concrete: Microstructure and Interfacial Transition Zone

S. Malazdrewicz(✉) and A. Chowaniec-Michalak

Wroclaw University of Science and Technology, Wybrzeże Wyspiańskiego 27, 50-370 Wroclaw, Poland
seweryn.malazdrewicz@pwr.edu.pl

Abstract. According to the European Construction Products Regulation, the use of recyclable and recycled materials is encouraged over conventional building materials. However, only about 20% of recycled mineral construction materials are currently utilized as aggregates, primarily in road construction and earthworks. Meanwhile, Eurostat data indicate that approximately 35% of buildings in the European Union (EU) are over 50 years old. In space-constrained Western European countries, the demolition of aging prefabricated concrete buildings is increasingly common, generating substantial amounts of waste that could be repurposed as recycled coarse aggregates (RCA) in concrete production. One of the main barriers to wider RCA usage is the weak interfacial transition zone (ITZ) between the adhered old mortar and the new cement paste. To investigate this issue, RCA was obtained from a prefabricated building constructed in the 1960s in the former German Democratic Republic. Three concrete mixtures were prepared with varying RCA content and analyzed using scanning electron microscopy (SEM). This study investigates the porosity, hydration products, and ITZ characteristics of RCA in new concrete and assesses its suitability for structural applications.

Keywords: Recycled Coarse Aggregate · Recycled Aggregate Concrete · Interfacial Transition Zone · Construction and Demolition Waste · Large Panel System Buildings

1 Introduction

Construction and demolition waste (CDW) accounts for more than one-third of all waste generated in the European Union [1]. Some components of CDW possess high resource value and technologies for waste separation and recovery are well established, but still recycling and material recovery rates vary significantly across EU member states—from below 10% to over 90% (see Fig. 1). Although the recycling of mineral construction waste is encouraged, recycled concrete is often used only for low-grade applications [2]. One of the challenges limiting its wider use in structural concrete is the uncertain quality of recycled coarse aggregates (RCA), particularly within the interfacial transition zone (ITZ) between old and new cementitious phases [3] (see Fig. 2). This

L. F. M. da Silva and R. D. Adams (Eds.): AB 2025, PEM, pp. 204–214, 2026.
https://doi.org/10.1007/978-3-032-19022-2_12

study explores the reuse of RCA sourced from demolished Large-Panel System (LPS) buildings—standardized prefabricated structures widely built across Europe in the mid-20th century. The consistent quality of LPS concrete may offer improved predictability for RCA-based concrete production [4]. Due to increasing pressure to reduce the carbon footprint of construction, incorporating recycled materials into concrete mixes has become a central topic in sustainable engineering. Among various recycled sources, RCA from prefabricated structures offers a unique advantage—predictable quality and uniform microstructure, stemming from standardized production methods used decades ago. Another critical aspect is the bond between RCA and new cement paste, which directly influences the mechanical performance of the resulting composite. Understanding and improving adhesion within the interfacial transition zone is therefore essential for enabling broader structural use of RCA.

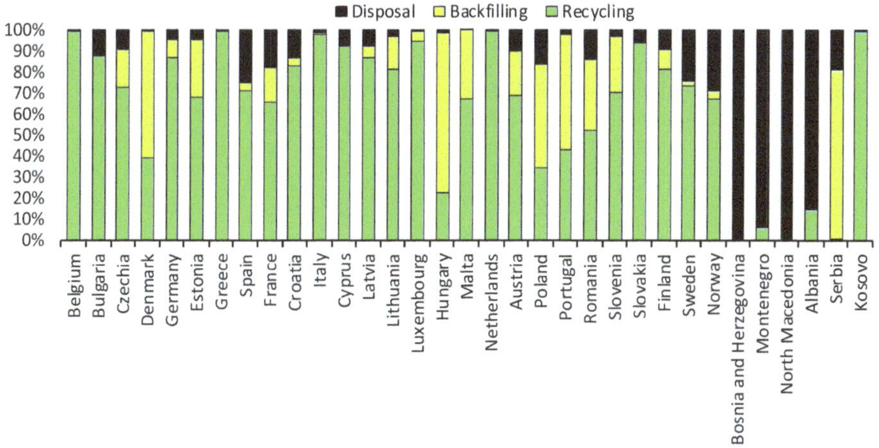

Fig. 1. Construction and Demolition Waste's mineral fraction treatment in 2022 (Eurostat, 2025)

Fig. 2. RCA characteristics: a) concrete waste before crushing b) possible ITZs in RAC

2 Materials and Methods

2.1 Materials

RCA was extracted from a 1960s LPS residential building in the former East Germany. Three concrete mixes were designed, replacing 10%, 20%, and 30% of natural coarse aggregate (NCA) with RCA. All mixes used CEM I 52.5 R cement, natural fine aggregate (0.25–2 mm), and coarse aggregate (4–8 mm), fly ash as a mineral additive, and Sika Viscocrete 2030HE as a superplasticizer. After 28 days of curing, samples were prepared for microscopic examination. Selected specimens underwent scanning electron microscopy (SEM) to observe porosity and hydration in the aggregate and matrix (S1, S2, and S3, respectively; see Table 1). Water from the municipal supply, meeting with PN-EN 1008, was used.

Table 1. Mix composition of concrete series with varying RCA content

Component	S1 [kg/m^3]	S2 [kg/m^3]	S3 [kg/m^3]
CEM I 52.5 R	475	475	475
Water	173	173	173
Fly ash	125	125	125
Superplasticizer	19.25	19.25	19.25
Natural coarse aggregate	1000	1000	1000
Natural fine aggregate	850	850	850
Recycled coarse aggregate	10%	20%	30%

2.2 Sample Preparation

Concrete was cast into 150 mm cubic moulds. After 24 h, the samples were demoulded and placed in water at a temperature of 20 ± 2 °C to cure for 28 days. From the matured concrete, specimens were cut for SEM analysis.

2.3 Microstructural Analysis

SEM analysis was conducted on NCA, RCA, and concrete specimens. Sample preparation for SEM analysis included: grinding with sandpapers of 320–1200 grit, polishing with diamond paste of 6 μm and 1 μm grain size, applying a copper path (to prevent surface charging and ensure electrical conductivity), and coating with a conductive carbon layer. The carbon layer, with a thickness of 30–40 nm, was applied under high vacuum (2×10^{-5} Pa) by thermal spraying using graphite electrodes. Microstructural images of the samples were taken using a JEOL JSM-6610A SEM with a conventional tungsten filament, in material contrast mode, using a backscattered electron detector (BSE). An accelerating voltage of 20 kV and a beam current of 39 nA were applied. Images were taken at 200× magnification in representative zones of each sample and in the ITZ using the method described in [5, 6]. For each sample, a single image was analysed. First, SEM images were imported into ImageJ software (National Institutes of Health, USA), where a grayscale histogram was analysed. Then, a cumulative histogram was generated, on the basis of which threshold values were determined using the point of intersection of two tangents to the curve at its inflection points. All pixels with values from 0 to the first threshold were considered pores, and the subsequent ranges were assigned to hydration products including C-S-H gel (HP), portlandite (CH), and unhydrated cement grains (AC). Portlandite was not always clearly distinguishable as a separate histogram range. Based on this, ImageJ software generated images of individual phase components as black pixels (0 value). Using Wolfram Mathematica software, the pixel content in each image was calculated, which allowed the porosity of the analysed sample and the content of other phases to be determined. The resulting images were combined with colour overlays.

3 Results

3.1 Porosity of Natural and Recycled Coarse Aggregates

Figure 3 shows microstructures of NCA and RCA at 20× magnification. NCA is characterized by a more compact structure compared to RCA, which exhibits a more irregular shape and irregular voids resulting from mechanical crushing. Phase contrast indicates the presence of many different components with varied chemical composition. NCA exhibited a uniform, dense structure with a higher proportion of visible micropores. RCA showed a more complex structure, with visible boundaries between old cement paste and the original aggregate. Surprisingly, RCA exhibited lower overall porosity, likely due to the highly matured cement paste of the original material.

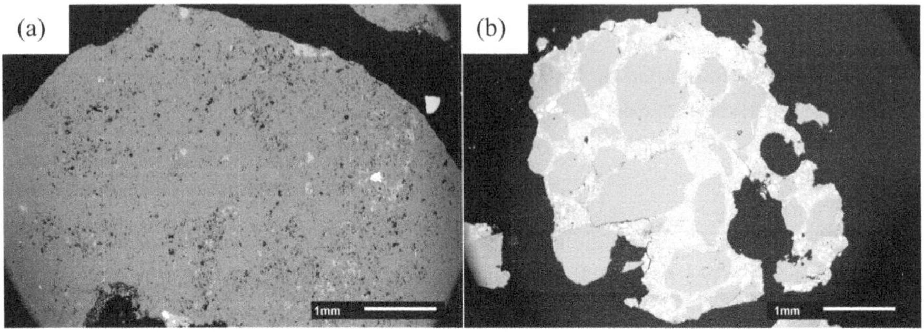

Fig. 3. Microstructure of: a) NCA, b) RCA, magnification 20×

In Fig. 4a to d, the porosity of NCA is shown, while in Fig. 4e to h the porosity of RCA aggregate is presented. The analysis was carried out based on SEM images taken at 200x magnification, for representative fragments of aggregate grains. Pores were identified as pixels with a value of 0 (black) on images of the samples previously converted to binary form. The NCA was characterized by a porosity of 11.0 and 12.0%, and in both presented images, micropores were visible, distributed throughout the entire cross-section of the grain. In the case of RCA, the porosity values were 3.4% and 4.7%. Clearly defined boundaries were observed between the remnants of the cement paste and the original aggregate, which constituted the original core of the RCA. In this original aggregate (being part of the RCA), pores were significantly less numerous compared to Fig. 4b and d. In the RCA aggregate, individual voids and air pockets were visible; however, their volume share was small. The average difference in porosity between the natural aggregate and RCA exceeded 7 percentage points, which indicates a clearly lower porosity of the recycled aggregate in the analysed samples.

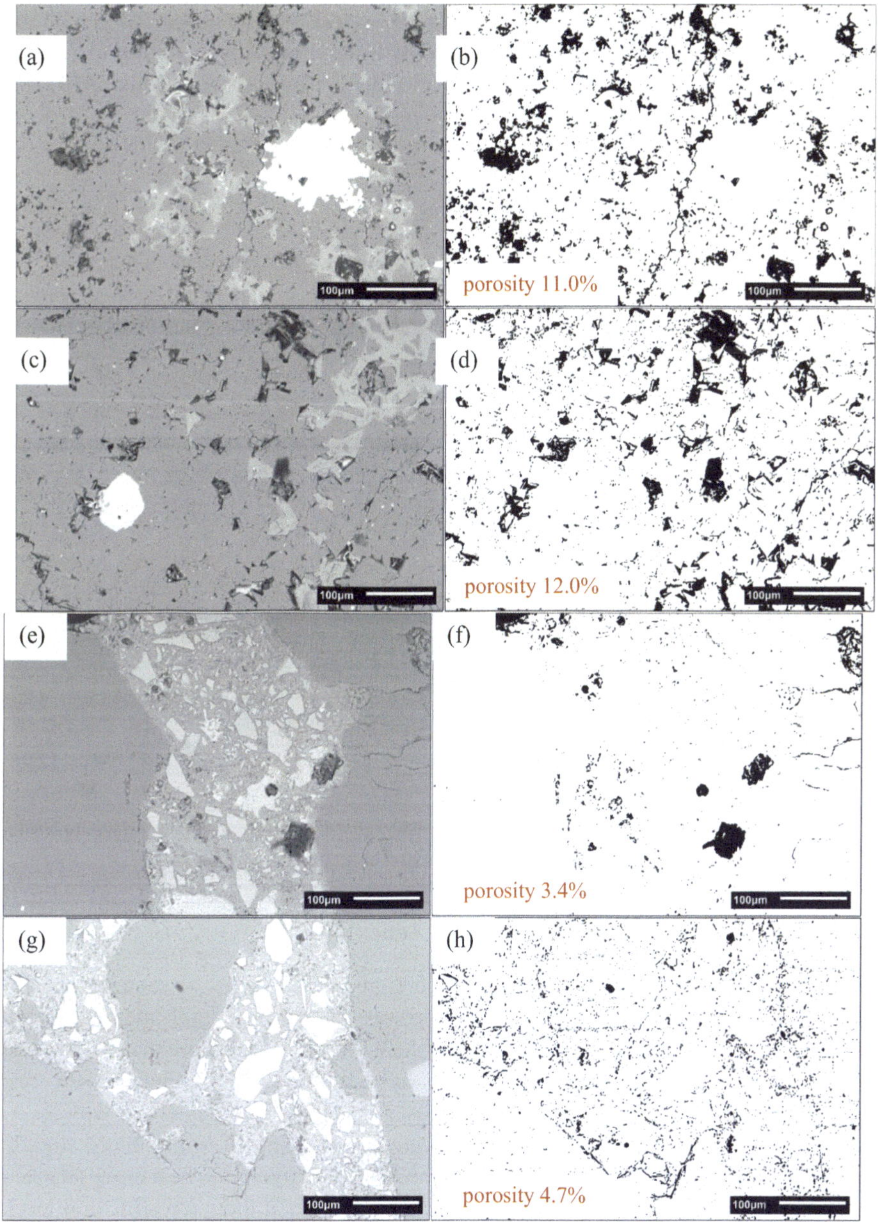

Fig. 4. SEM and pore images at 200× magnification: a) b) c) d) of NCA e) f) g) h) of RCA

3.2 Porosity of Natural and Recycled Coarse Aggregates

Figure 5 shows the grayscale histogram after digitally removing the original aggregate, which made it possible to isolate the cement matrix area. Distinct peaks in the histogram confirm the presence of well-differentiated phases in the analysed region. Based on the analysis of the histogram and cumulative histogram, tonal ranges corresponding to individual phases were determined: porosity, hydration products including C-S-H gel (HP), portlandite, and unhydrated cement grains (AC). The degree of hydration (α) was estimated based on the Powers and Brownyard model [7], using Eq. 1, where F_{hp} denotes hydration products (HP + CH), and F_{up} denotes unhydrated cement grains (AC).

$$\alpha = \left(F_{hp}/2.1\right) / \left[\left(F_{hp}/2.1\right) + F_{up}\right] (-) \tag{1}$$

All samples showed well-formed hydration products. The internal cement paste of the RCA had a high degree of hydration, attributed to its long service life. The fresh cement matrix in RCA-based concrete also showed good hydration performance, particularly in the mixture with 30% RCA (Fig. 5).

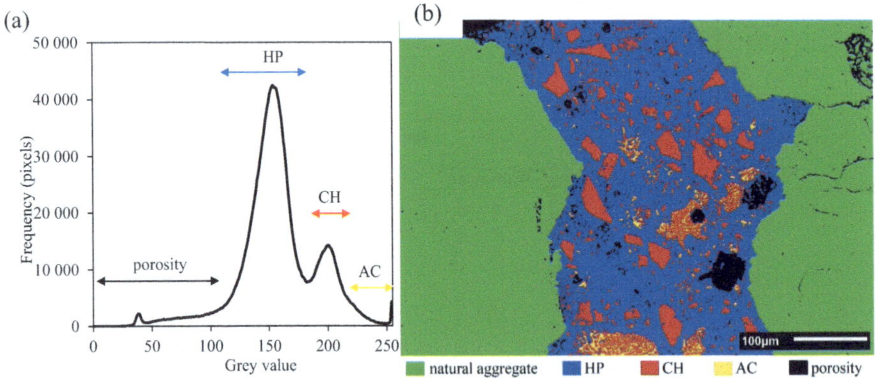

Fig. 5. (a) Grayscale histogram; (b) SEM image with marked phases: natural aggregate, porosity, hydration products (HP), portlandite (CH), unhydrated cement grains (AC)

3.3 Analysis of the Interfacial Transition Zone (ITZ)

In the next step, ITZ regions were extracted from SEM images of samples S1, S2, and S3. For each sample, a 50 μm-wide strip was defined in ImageJ, measured from the aggregate boundary into the cement matrix. Within this area, pore distribution and phase segmentation were carried out (Fig. 6). In samples S1 and S3, the analysed zone bordered RCA, while in sample S2 the ITZ was extracted along the interface with NCA. All images were analysed at 200 × magnification. Although RCA introduced slightly higher porosity in the ITZ, it also supported more advanced hydration near the aggregate–matrix interface, which may positively influence adhesion at the microstructural level. The presence of old, hydrated cement paste may help retain moisture locally, improving ongoing hydration of the new matrix. Overall, the ITZ in RCA mixes was more porous but contained fewer unhydrated cement particles than that in NCA mixes.

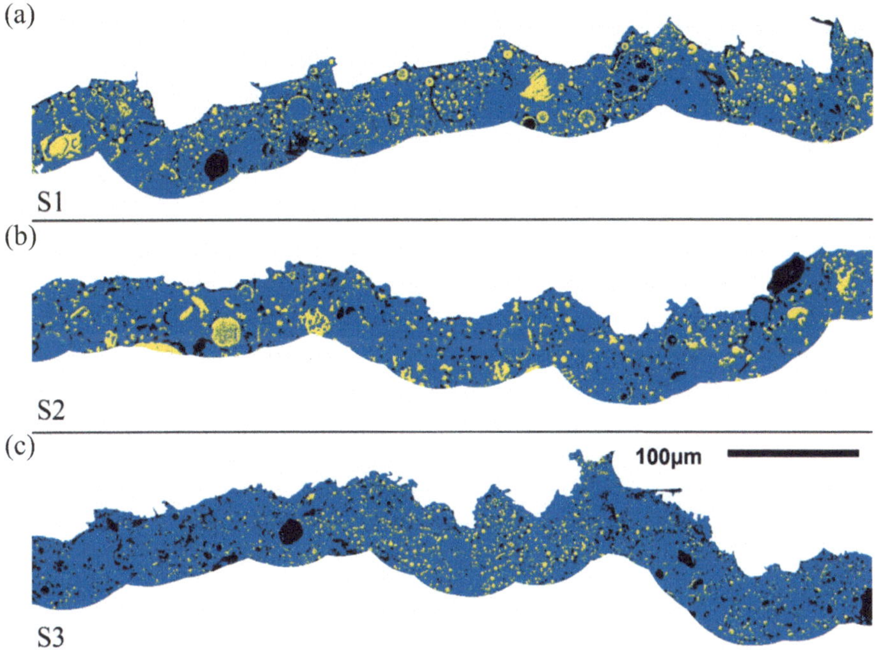

Fig. 6. Phase maps of the 50 µm-wide interfacial transition zone (ITZ) in samples S1 (a), S2 (b), and S3 (c). Phases: porosity (black), HP (blue), and AC (yellow)

Figure 7 shows the phase distribution along the length of the interfacial transition zone. The graphs illustrate the local content of porosity, hydration products (HP), and unhydrated cement (AC) along the aggregate–matrix interface. The results indicate microstructural differences in the ITZ depending on the type of adjacent aggregate. For example, sample S3 (RCA) showed relatively high porosity (average 9.4%) and the lowest AC content (4.9%), while sample S2 (NCA) had the highest amount of unhydrated cement (10.3%) and the lowest porosity (6.2%). These results suggest that using recycled aggregate may lead to a more porous ITZ, but at the same time support the hydration process.

To better highlight the differences, Fig. 8 presents the average phase composition within the ITZ for all three samples. Sample S2, containing NCA, had a moderate share of hydration products (83.5%) and AC (10.4%). In contrast, sample S3 showed the most advanced hydration process (85.7%) and the lowest amount of unhydrated cement (4.9%). These trends may be related to differences in water absorption and surface roughness between RCA and NCA, which affect the local water-to-cement ratio and the progress of hydration reactions within the ITZ. Overall, the results confirm that the interfacial transition zone is influenced by the type of coarse aggregate used. RCA, despite increasing local porosity, may support more effective hydration near the aggregate–matrix interface.

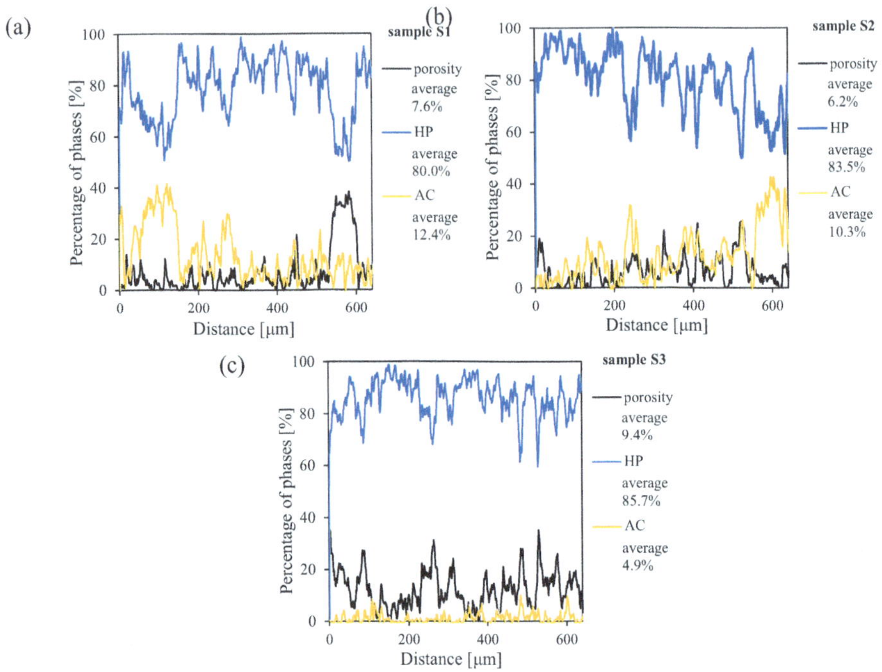

Fig. 7. Phase distribution along the interfacial transition zone (ITZ) with a width of 50 μm in samples S1 (a), S2 (b), and S3 (c)

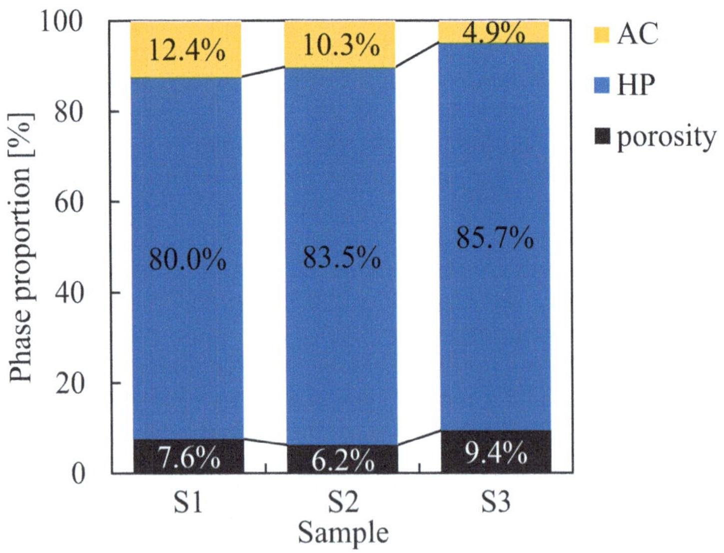

Fig. 8. Average phase composition within the 50 μm-wide ITZ for samples S1, S2, and S3

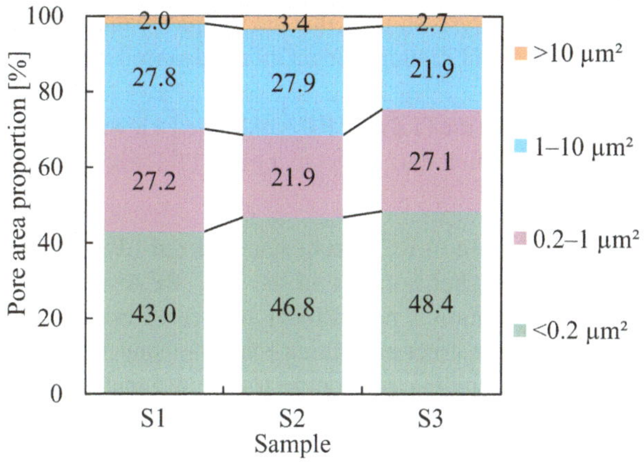

Fig. 9. Proportional pore area distribution within the ITZ classified by pore size ranges for samples S1–S3

Figure 9 presents the distribution of pore sizes in the ITZ for samples S1–S3, grouped into four pore size ranges. In all samples, the smallest pores (<0.2 μm^2) dominated, accounting for 43.0% (S1), 46.8% (S2), and 48.4% (S3) of the total pore area, respectively. The greatest variation was observed in the intermediate pore range (0.2–1 μm^2), with the highest proportion in sample S1 (27.2%) and the lowest in sample S2 (21.9%). The contribution of the largest pores (>10 μm^2) was relatively low. These results suggest that the type of aggregate used may influence not only the overall porosity of the ITZ but also the distribution of pore sizes within this zone.

4 Conclusions

RCA from precast LPS buildings demonstrates favourable microstructural characteristics, including low porosity and high hydration in the old paste. The cement matrix in new RCA concrete showed no adverse effects, even at a 30% RCA replacement level. Despite a more porous ITZ, the hydration degree near RCA was often higher than near NCA, suggesting that the old adhered mortar may support better moisture retention and promote stronger adhesion at the interface. These findings support the feasibility of using RCA from LPS buildings in structural concrete, promoting sustainable construction practices. Based on the microstructural analysis, the following conclusions were drawn:

- Recycled coarse aggregate (RCA) exhibited lower total porosity than natural coarse aggregate but included visible remnants of old mortar.
- Increased RCA content in concrete was associated with higher porosity and a reduced share of hydration products in the cement matrix.
- The interfacial transition zone (ITZ) consistently exhibited higher porosity than the surrounding matrix, likely due to the irregular shape of aggregate grains and limited paste encapsulation.

- The phase distribution analysis showed that the sample containing RCA (S3) had the highest share of HP in the ITZ, suggesting a more advanced degree of hydration near the RCA interface.
- Despite its higher porosity, the ITZ near RCA exhibited a lower content of unhydrated cement particles (AC), which may indicate a more effective hydration process than in ITZ with NCA.

This study supports the use of RCA from standardized LPS buildings in structural concrete. Despite slightly higher porosity in the ITZ, RCA showed a well-hydrated structure and promoted hydration in new cement matrices. Results confirm its potential to contribute to sustainable construction without compromising microstructural integrity. Future work will focus on expanding the scope of testing and analysis to comprehensively evaluate the performance of RCA in concrete.

Funding. This research was funded from a project supported by the National Centre of Science, Poland [Grant no. 2020/39/O/ST8/01217 "Experimental evaluation of the fundamental properties of self-compacting concrete made of recycled coarse aggregate sourced from the demolition of large panel system buildings (RECSCCUE)"].

References

1. European Commission. Directorate-General for Environment: Energy, Climate Change, Environment (2025). https://environment.ec.europa.eu/. Accessed 7 Aug 2025
2. Pacheco, J., de Brito, J., Lamperti Tornaghi, M.: Use of recycled aggregates in concrete: opportunities for upscaling in Europe. Publications Office of the European Union, Luxembourg (2023). https://doi.org/10.2760/144802
3. Dettenborn, T., Harju, I., Forsman, J., Lehtonen, K., Ahlqvist, E., Kurki, T.: Recycling of concrete from demolished bridges and other engineered structures. In: IOP Conf. Ser.: Earth Environ. Sci. **710**, 012064 (2021). https://doi.org/10.1088/1755-1315/710/1/012064
4. Malazdrewicz, S., Ostrowski, K.A., Sadowski, Ł: Self-compacting concrete with recycled coarse aggregates from concrete construction and demolition waste–Current state-of-the-art and perspectives. Constr. Build. Mater. **370**, 130702 (2023). https://doi.org/10.1016/j.conbuildmat.2023.130702
5. Wong, H.S., Head, M.K., Buenfeld, N.R.: Pore segmentation of cement-based materials from backscattered electron images. Cem. Concr. Res. **36**, 1083–1090 (2006). https://doi.org/10.1016/j.cemconres.2006.03.007
6. Scrivener, K.L., et al.: TC 238-SCM: Hydration and microstructure of concrete with SCMs: State of the art on methods to determine degree of reaction of SCMs. Mater. Struct. **48**, 835–862 (2015). https://doi.org/10.1617/s11527-014-0527-z
7. Powers, T.C., Brownyard, T.L.: Studies of the physical properties of hardened Portland cement paste. J. Am. Concr. Inst. Proc. **43**(9), 549–602 (1947)

Author Index

L. F. M. da Silva and R. D. Adams (Eds.): AB 2025, PEM, p. 215, 2026.
https://doi.org/10.1007/978-3-032-19022-2

The manufacturer's authorised representative in the EU is Springer Nature Customer Service Centre GmbH, Europaplatz 3, 69115 Heidelberg, Germany. If you have any concerns regarding our products, please contact ProductSafety@springernature.com

Printed and bound by CPI Group (UK) Ltd, Croydon, CR0 4YY
07/10/2026
02229349-0003